T0229806

CAMBRIDGE TEXTS IN THE
HISTORY OF PHILOSOPHY

———

ROBERT BOYLE
*A Free Enquiry into the Vulgarly
Received Notion of Nature*

CAMBRIDGE TEXTS IN THE HISTORY OF PHILOSOPHY

Series editors

KARL AMERIKS
Professor of Philosophy at the University of Notre Dame

DESMOND M. CLARKE
Professor of Philosophy at University College Cork

The main objective of Cambridge Texts in the History of Philosophy is to expand the range, variety and quality of texts in the history of philosophy which are available in English. The series includes texts by familiar names (such as Descartes and Kant) and also by less well-known authors. Wherever possible, texts are published in complete and unabridged form, and translations are specially commissioned for the series. Each volume contains a critical introduction together with a guide to further reading and any necessary glossaries and textual apparatus. The volumes are designed for student use at undergraduate and postgraduate level and will be of interest not only to students of philosophy, but also to a wider audience of readers in the history of science, the history of theology and the history of ideas.

For a list of titles published in the series, please see end of book.

ROBERT BOYLE

A Free Enquiry into the Vulgarly Received Notion of Nature

EDITED BY

EDWARD B. DAVIS

Messiah College, Grantham, Pennsylvania

MICHAEL HUNTER

Birkbeck College, University of London

CAMBRIDGE
UNIVERSITY PRESS

Published by the Press Syndicate of the University of Cambridge
The Pitt Building, Trumpington Street, Cambridge CB2 1RP
40 West 20th Street, New York, NY 10011–4211, USA
10 Stamford Road, Oakleigh, Melbourne 3166, Australia

First published 1996

A catalogue record for this book is available from the British Library

Library of Congress cataloguing in publication data

Robert Boyle: a free enquiry into the vulgarly received notion of nature
/ edited by Edward B. Davis, Michael Hunter.
p.   cm. – (Cambridge texts in the history of philosophy)
Includes bibliographical references.
1. Nature – Early works to 1800.
2. Science – History – 17th century – Early works to 1800.
3. Religion and science – History – 17th century – Early works to 1800.
I. Davis, Edward Bradford. 1953– .   II. Hunter, Michael Cyril William.   III. Series.
Q155.R62   1996
508–dc20   96-10997   CIP

ISBN 0 521 56100 0 hardback
ISBN 0 521 56796 3 paperback

Transferred to digital printing 2001

Contents

Acknowledgements

The following have assisted in the preparation of this edition. In our work on the Boyle Papers, we have been greatly helped by the library staff at the Royal Society. Work on the Latin edition of the text was carried out as part of the preparation of the forthcoming 'Pickering Masters' edition of *The Works of Robert Boyle*, supported by generous grants from the Leverhulme Trust and the Royal Society, and by the award to Edward B. Davis of a Mellon Fellowship at the University of Pennsylvania in 1991–2. A Scholarship Support Grant from Messiah College provided Professor Davis with a reduced teaching load to prepare the notes and finalise the modernised text, in which he was ably assisted by Meredith Fritz, an undergraduate student supported by the College's Scholar Intern Program. Desmond Clarke's extensive comments on the text have been invaluable, while the help of Lawrence Principe and Harold Cook has been indispensable in elucidating various of Boyle's allusions. In addition, the following have assisted in various ways: Clive Cheesman, Kate Fleet and Lesley Suckling.

Abbreviations

BP   Royal Society Boyle Papers

Lat.   Robert Boyle, *De Ipsa Natura, sive Libera in Receptam Naturae Notionem Disquisitio ad Amicum*, Latin translation by David Abercromby, London, 1687

MS   Royal Society Manuscripts

Works   *The Works of the Hon. Robert Boyle.* Edited by Thomas Birch, 2nd edn., 6 vols., London, 1772

Introduction

The scientific revolution of the seventeenth century was a turning point in Western thought. It is largely to this era that we owe the ethos of modern science – empirical, cumulative and deeply quantitative in its methods; iconoclastic in its view of previous intellectual traditions; and assertive of its ability not only to understand nature but also to control it. Perhaps most fundamental, however, was the adoption at this time of a new world view, involving a complete change in the way in which nature was conceived. By reviving and adapting the ancient atomist conception of nature, the mechanical philosophers of the seventeenth century challenged prevailing Aristotelian and Galenic notions, which typically depicted nature as a wise and benevolent being. Associated with such views were phrases like 'Nature does nothing in vain', 'Nature abhors a vacuum' or 'Nature is the wisest physician'. By contrast, thinkers such as René Descartes (1596–1650), Pierre Gassendi (1592–1655) and Robert Boyle (1627–91) held that the world was a vast, impersonal machine, incapable of acting consciously.

Theological issues were intimately and inextricably involved in this conceptual shift in ways that have only recently been fully appreciated. Traditionally, it was presumed that the acceptance of mechanical explanations eroded belief in divine providence and drove a wedge between theology and science. Although it is true that mechanistic science heightened some existing tensions within the doctrine of creation, it is equally true that many of its advocates thought that it was actually more consistent with biblical statements of divine sovereignty than older, non-mechanistic views – and that this was highly relevant to its value as a theory of nature. No one embodied this outlook better than

Boyle, the most famous mechanical philosopher of his age, and nowhere did he express himself more fully on these matters than in *A Free Enquiry into the Vulgarly Received Notion of Nature* (1686), one of his subtlest and most significant writings. In it, Boyle deployed his philosophical acumen, theological learning and experimental expertise to explore the implications of rival conceptions of the natural world, especially in questioning prevailing Aristotelian and Galenic notions.

For Boyle, it was inappropriate both theologically and scientifically to speak of 'Nature' doing anything at all. Instead, he argued for the superior intelligibility of the mechanistic view of the world that he had championed in his profuse earlier writings, a world made up of matter acting according to properties and powers given to it by God. Moreover, he claimed that such a view was closely tied to a proper conception of God's absolute power over the world, from which he saw the 'vulgar' view as detracting. By denying 'Nature' any wisdom of its own, the mechanical conception of nature located purpose where Boyle believed it belonged: over and behind nature, in the mind of a personal God, rather than in an impersonal semi-deity immanent within the world.

This introduction will set Boyle's work in context by dealing with its author and his intellectual background, with the antagonists whom he confronted in his book, with the history of the work's composition and publication and with its impact and legacy.

Robert Boyle

Robert Boyle was the leading British natural philosopher in the generation before Newton (throughout this introduction, we will use the words 'natural philosopher' and 'scientist' interchangeably, since we see them as effectively synonymous, although 'scientist' was not used before the nineteenth century). By birth he was the Honourable Robert Boyle: his father was Richard Boyle, Earl of Cork, one of the most powerful men in England before the Civil War. Following an education at home and at Eton College, Boyle spent the years from 1639 to 1644 travelling in France, Switzerland and Italy. The year after his return to England, he settled at Stalbridge in Dorset, where he had inherited an estate from his father.

Boyle's earliest writings were not on science but on the pursuit of morality and piety, intended to inculcate spiritual and ethical awareness into his dissipated peers. Only one of these was published during his lifetime, *Some Motives and Incentives to the Love of God* (1659; usually referred to as *Seraphic Love*), but many of the rest survive, and they combine a concern for stylistic elegance with a strong emphasis on the moral import of knowledge: though, as we shall see, the former is only vestigially in evidence in his later writings, the latter seems to have remained with him throughout his life.

Already during this moralistic phase – and even during his European tour – Boyle had taken some interest in the marked changes in understanding of the workings of the universe that had occurred during the previous century, notably the astronomical revolution associated with Nicolaus Copernicus (1473–1543), Johannes Kepler (1571–1630) and Galileo Galilei (1564–1642), to which he refers in the text below. As far as his own commitment to experimental science is concerned, a major turning-point occurred around 1650. In 1649 he successfully set up a laboratory at his house at Stalbridge, and he seems to have been transfixed by the experience. 'Vulcan has so transported and bewitched me, that ... [I] fancy my laboratory a kind of Elysium', he told his sister, Katherine, Lady Ranelagh,[1] alluding to the furnaces that he had erected by way of reference to the classical god of fire: from this time onwards, his moralistic writing was eclipsed by an increasing fascination by experimentally based knowledge. Moreover, his valuation of it was from the outset linked to what he saw as its superiority to other forms of knowledge from the point of view of the support it could provide for religion.

Boyle's discovery of experiment was accompanied by a parallel discovery of erudition. At the same time as he was building his laboratory at his Stalbridge home, he was also acquiring a knowledge of the ancient languages, Hebrew, Caldaic, Syriac and Arabic, and a familiarity with the vast learning that scholars in his period had accumulated concerning the religious and other ideas of the ancient world. His activity in this sphere reached its peak in the early 1650s, but it left him with an expertise that he was able to draw on later, for instance in the excursus in section IV of his *Free Enquiry* in which he considered the link between the personification of nature and ancient polytheism.

---

[1] Boyle to Lady Ranelagh, 31 August 1649, *Works*, vol. 6, pp. 49–50.

Late in 1655 or early in 1656 Boyle moved to Oxford, where he joined the group of natural philosophers there assembled under the aegis of the divine, John Wilkins (1614–72), which included such luminaries as Christopher Wren (1632–1723), Seth Ward (1617–89) and Robert Hooke (1635–1703). This was the liveliest group of scientific intellectuals then existing in England, and Boyle's association with it further intensified his commitment to science. The Oxford group prefigured the Royal Society, founded in 1660, both in the range of its members' interests, and in their attempt to graft together the empirical approach to nature enunciated earlier in the century by Francis Bacon (1561–1626) with the novel conceptualisation of the natural world put forward by continental thinkers like Gassendi and Descartes. It was evidently at this time that Boyle became thoroughly acquainted with Gassendi's revival of atomism, the ancient conception of matter as comprised of indivisible particles, and with the mechanistic theory of the world associated especially with Descartes. Towards the latter, in particular, Boyle adopted a slightly ambivalent stance which he retained throughout his life, and to which he gives expression in the text below.

Boyle's Oxford years saw an extraordinary burst of activity on his part, since he now began or completed a whole series of books on different aspects of natural philosophy. It was at this time that he wrote the bulk of his *The Usefulness of Natural Philosophy*, protesting the value of improved understanding of the natural world not only in its own right but also for religious enlightenment and for the affairs of life. His *Certain Physiological Essays* explored and vindicated the value of experiment, and sought experimentally to demonstrate the validity of a mechanistic theory of matter, to which he gave the name 'corpuscularianism' to avoid the irreligious overtones that atomism had inherited from classical antiquity. He also began work at this time on experimental studies of colour and of cold, and he wrote his most famous book, *The Sceptical Chymist*, a somewhat discursive assessment of the theories both of the Aristotelians and of his chief predecessors as experimenters, the chemists. Lastly, his *New Experiments Physico-Mechanical, Touching the Spring of the Air*, written in a relatively short period in 1659–60 and published soon after it was finished, recounted experiments which Boyle carried out using a vacuum chamber or air pump: this enabled him to illustrate the characteristics and functions of the air by studying the effects of its withdrawal on flame, light and living creatures.

New Experiments inaugurated a profuse programme of publication on Boyle's part. All of the works referred to in the previous paragraph came out between 1660 and 1665, and they established his reputation as the leading natural philosopher of his day. Most of them appeared in Latin editions as well as English – in many cases published under Boyle's own auspices at London or Oxford – and this ensured that his fame spread through Europe. Boyle's production of treatises dealing with a range of scientific topics continued for the rest of his life, while later in the 1660s he supplemented this by publishing his findings in the form of articles in the *Philosophical Transactions* of the newly founded Royal Society. This was edited by Henry Oldenburg (1618?–77), first secretary of the Society and a protégé of Boyle's, and he used this organ to promote Boyle's ideas, thus further contributing to his renown.

Boyle was from the outset closely associated with the Royal Society, the first modern scientific institution. He was present at the Society's inaugural meeting on 28 November 1660, and he attended its meetings regularly in its early years, participating in the profuse programme of experiment and discussion that characterised its early activities. The Society's objectives were encapsulated in a letter from Oldenburg to an overseas correspondent in 1668: 'it aims at the improvement of all useful sciences and arts, not by mere speculations, but by exact and fruitful observations and experiments'.[2] Its hallmark was a commitment to the Baconian ideal of empirical investigation, which was applied to a range of subjects that constituted what was emerging at the time as 'science' in a recognisably modern sense, centring on natural and mechanical problems but extending through the life sciences towards medicine and through chemistry and applied mathematics towards technology. The Society quickly established itself as the most influential grouping in European natural philosophy in its day, and this was due not least to the extent to which leading scientific figures like Boyle were associated with it. Even after 1670, when ill health and other factors meant that Boyle ceased to attend the Society's meetings on a regular basis, he continued to identify himself with the institution: thus he describes himself as 'Fellow of the Royal Society' on the title-page of his *Free Enquiry*. More than almost any other individual, he was crucial to

---

[2] Oldenburg to Richard Norwood, 10 Feb. 1668, in A. R. and M. B. Hall (eds.), *The Correspondence of Henry Oldenburg* (13 vols., Madison, Milwaukee and London, 1965–86), vol. 4, p. 168.

the Society's image, since his profuse writings seemed to epitomise the ideal of painstaking, empirical investigation of the natural world which was the goal of the Society and of the new science in England in this formative phase.

In addition to his assiduous attempts to elucidate the workings of nature by experiment, Boyle also wrote extensively about the philosophical corollaries and limitations of knowledge about the natural world, and about the mutual relationship of science and religion. Indeed, his writings on such subjects can claim a significance almost equal to that of his experimental work. As has already been noted, it was its apologetic role in relation to religion that had attracted Boyle to natural philosophy in the first place – even if he also found experimental knowledge absorbing in its own right – and this apologetic concern continued thereafter. Like many contemporaries, Boyle was preoccupied by the threat of an ill-defined 'atheism', and he was therefore constantly anxious to oppose trends in ideas which seemed to abet this, and to champion intellectual methods which appeared to offset them.

First and foremost among these was experiment, for Boyle was convinced that, properly understood, the new, mechanistic science that he championed was an invincible ally for religion – indeed, more so than rival views of nature, including the 'vulgar notion' attacked in this work. Equally worrying was the threat of materialism: here the dangers seemed to be exemplified by Boyle's English contemporary, the philosopher Thomas Hobbes (1588–1679), to the perniciousness of whose ideas Boyle seems first to have been alerted in the mid-1650s. Largely because of his hostility to the religious implications of Hobbes's ideas, Boyle allowed himself to become involved in an intense debate with Hobbes in the 1660s which continued into the 1670s.

Such debates served the ancillary purpose of distancing Boyle's own mechanistic science from materialism like Hobbes's, and vindicating it from the imputation that it, too, might lead in an atheistic direction. Boyle was only too aware that some of his contemporaries saw the new science as a threat to religion, due to an association of ideas inherited from classical antiquity. He sought to show that, on the contrary, his own understanding of nature was not only compatible with religion but positively supportive of an intense appreciation of God's goodness and power. This was a subject on which he wrote at length, reaching a climax with his *The Christian Virtuoso: Shewing, That by Being addicted to Experimental Philosophy, a man is rather assisted than indisposed to be a good Christian* (1690).

Boyle had undergone a conversion experience during his adolescence, when like Martin Luther he was terrified by a thunderstorm and vowed to live piously if his fears should be unrealised. He kept his vow, for the depth and sincerity of his religiosity were often remarked on by those who knew him. Closely tied to Boyle's spirituality was his profound belief in an omnipotent creator who had made the world freely, not out of necessity; thus the laws of nature could not be found a priori from first principles, but had to be discovered from the works of creation. This particular theological orientation is known as 'voluntarism', for it emphasises the voluntary choices of a free creator, rather than ways in which God's choices were determined by the dictates of reason, either human or divine. Indeed for Boyle, one of the most attractive features of the mechanical philosophy was the extent to which it removed mediating influences between God and the world, thereby preserving God's sovereignty more clearly than the 'vulgar' notion of nature which, in Boyle's opinion, elevated nature to the status of a semi-deity. Boyle's voluntarism is most evident in his strong supernaturalism, according to which God had in the past and might at any time in the future suspend the ordinary course of nature, acting in special ways to achieve particular ends. Thus, Boyle was fascinated by evidence from miracles and fulfilled prophecies, which appeared to vindicate such a role in the world on God's part.

As part of his practical Christianity, Boyle was deeply interested in the power of healing, writing extensively on medical matters. The longest single section of his *Usefulness of Natural Philosophy* dealt with medicine, and he returned to similar themes in publications in the last decade of his life. In these books, Boyle explored the medical spin-offs of his natural philosophical research: though he planned a work that would have gone beyond this to criticise the orthodox medical practice of his day, he failed to publish this, not least because he did not wish to be accused of trespassing in an area where he lacked professional qualification.[3] As will be seen, though he dealt at length with medical

---

[3] See Michael Hunter, 'Boyle versus the Galenists: a Suppressed Critique of Seventeenth-century Medical Practice and its Significance', forthcoming. See also B. B. Kaplan, *'Divulging of Useful Truths in Physick': The Medical Agenda of Robert Boyle* (Baltimore, Johns Hopkins Press, 1993).

issues in the *Free Enquiry*, he is here similarly apologetic concerning his excursion into the medical sphere. Nevertheless, these passages illustrate his deep knowledge of contemporary medical thought and practice.

Taken together, Boyle's writings on scientific, philosophical, medical and religious topics comprise some forty books. In his early years, before he turned to science, his self-consciousness as a writer had been reflected in elegantly turned sentences, elaborate metaphors and other rhetorical flourishes. Some remnants of these stylistic pretensions survived into his later years, as will be seen from the well-chosen metaphors and other devices that appear in the text below. In general, however, Boyle's mature style is highly convoluted and digressive, and it is sometimes hard to follow. In part this is because, from the early 1650s onwards, a serious illness which affected his eyes meant that Boyle never wrote more than short passages himself, instead dictating his words to amanuenses. In addition, his scrupulous concern for accuracy meant that he constantly felt the need to qualify points – often thereby obscuring his meaning – while he never tired of adding extra illustrative material. The result is often not a pleasure to read, since Boyle's seemingly irresistible tendency to digress – sometimes even within a digression – was combined with a remarkable inability to bring a sentence to a suitable close. It is not uncommon to find a page of text which comprises three or four independent clauses, half a dozen dependent clauses, one or two parenthetical expressions and more commas than one would care to count – but not a single full stop! Yet it may be argued that this very convolution brings us close to Boyle as a thinker, struggling to do justice to ideas that were often highly complex about the relationship between God and nature, or the strengths and weaknesses of rival philosophical views.

Boyle's antagonists

Boyle is often rather inexplicit in naming those with whom he disagreed, especially if they were contemporary authors. Indeed, his unhelpful vagueness in identifying those whose ideas he was attacking is responsible for one of the outstanding problems of interpreting his *Free Enquiry* – his reference on p. 47 to the 'sect' of Christian philosophers who had revived ancient views identifying God with nature. This has been the subject of widely divergent interpretation, largely because Boyle's wording is so opaque.

What Boyle is more explicit about are the broad categories of opinion with which he disagreed, and it is these that we will deal with here. First, there are the materialist infidels, of whom Hobbes is named in connection with his view of the corporeality of God, though in general Boyle's references are limited to such ancient authors as Epicurus. Such thinkers are generally held up as espousing extreme (and, implicitly, absurd) materialistic positions, believing that the world had originated by chance and that no design underlay it at all. The utter incredibility of materialism was not a topic that Boyle pursued at length in the work below, but he devoted more space to it in his *Disquisition about the Final Causes of Natural Things* (1688), which he wrote in parallel with the *Free Enquiry* and which he may have seen as complementing it.

The threat that seemed to Boyle more severe both because of its greater plausibility and of its pervasiveness in the thought of his day was the system of ideas based on the texts of the ancient Greek philosopher, Aristotle. Having been christianised by Aquinas and others in the Middle Ages, Aristotelianism had come to form the basis of the university curriculum, which it still dominated in Boyle's day. It was in this sense of 'commonplace' that Boyle used the term 'vulgar' to describe the point of view that he was discussing. Though he saw it as having resonances in everyday observations and sentiments (as indeed Aristotelianism did), it was among the learned that he expected his chief antagonists to be ranged. These are described as 'Peripatetics' (alluding to Aristotle's habit of teaching while walking around the Lyceum at Athens) or 'school philosophers' or 'scholastics' (denoting the academic milieu in which such ideas flourished).

According to Aristotle, the universe was a highly ordered place that displayed, not an order imposed from the outside by a transcendent creator, but a functional teleology immanent within nature, according to which the 'nature' or 'essence' of a body determined its properties. Aristotelian cosmology offers a clear example. All terrestrial things were made of four elements – earth, water, air and fire – each of which had a place where it properly belonged according to its nature, and to which it sought to return when removed from it: earth at the centre of the universe (which coincided with the centre of the earth), water on the earth's surface, air above the earth's surface and fire above air, just below the sphere of the moon. By contrast the heavens, which included the moon, the sun, the fixed stars and the planets, consisted of perfect spheres of ether, a fifth element utterly unlike the four terrestrial elements. Within this cosmological scheme, the 'natural' motion of an object was determined by the nature of the element of which it was predominantly composed. For example, a rock falls 'naturally' to the ground because it is made mostly of the element earth, and it is the 'nature' of earth to seek its 'natural' place at the centre of the universe. In seeking its natural place, a rock moves downward as far as it can, stopping only when it collides with the ground. The 'natural' motion of the ethereal spheres, on the other hand, was to rotate on their axes, thereby carrying the planets round the earth.

The fall of the rock was understood further as a change of place analogous to a change of colour or size in a ripening apple; local motion was just one specific kind of change from potentiality to actuality, another key Aristotelian concept. For example, a seed is not an actual plant, but a potential plant; its growth is a process of becoming an actual plant, involving a certain type of motion towards a definite end: the actual state of being a mature plant. Just as a seed fulfils its nature by becoming a plant, so a rock fulfils its nature by moving towards its natural place. In a very real sense, then, Aristotelian physics was fundamentally biological in orientation: the world was essentially a vast organism, with the behaviour of its parts determined by their natures and purposes.

If, however, the same rock were to be thrown upward, its motion was said to be 'unnatural' or 'violent', since it was moving away from its natural place, although natural motion would be restored once the rock ceased rising and began to fall. Likewise, machines and other 'artificial' devices were capable of forcing bodies to move 'unnaturally'. Medieval and Renaissance Aristotelians called this kind of artificially produced motion 'preternatural', because it was thought to be outside the course of nature. The preternatural also included events with causes which were unnatural, but not supernatural, such as the birth of a two-headed animal. Boyle examines the use of these terms in section VI of the present work.

Another feature of scholasticism was the explanation of natural events in terms of forms, qualities and faculties: these were descriptions of what ordinary, common-sense observations tell us about the properties of a body. 'Qualities' were the attributes of a body as well as the 'substantial forms' or 'faculties' that caused the attributes. Thus, for example, certain medicines were said to possess an 'aperitive faculty' (derived from the Latin verb 'aperīre', 'to open') if they could 'open up' the body in some manner, such as purgatives or diuretics. Once again, we see how a particular activity is explained in terms of the 'nature' of a body. In what is really a joke at his opponents' expense, Boyle notes in the final section of the treatise that the schoolmen would describe the ability of a key to open a door – for him, an obvious example of a mechanical property – as an aperitive faculty!

Related to forms and qualities was the distinction between substance and accident, which Boyle examines near the start of section VIII, where he asks whether the 'vulgar' notion of nature itself is a substance or an accident. According to Aristotelian doctrine, the 'substance' of a thing denotes the 'stuff' it consists of, the material or immaterial substratum separated conceptually from the properties or 'accidents' of the thing. For example, the substance of wine is the liquid body itself, which was regarded as consisting mostly of water. The accidents of wine include its taste, smell and colour. Boyle notes that most proponents of the 'vulgar' notion regarded nature as a substance; he wants to know what sort of substance it is, corporeal or incorporeal.

The differences between Aristotelianism and the various mechanical philosophies of the seventeenth century have sometimes been misstated. It is not true that Aristotelians downplayed or ignored observations. In fact, they relied heavily on direct sense experience at the expense of abstract theorisation, as Galileo realised. Indeed, many Aristotelians believed that scientific explanations ought not to go beyond the manifest properties of bodies to speculate about 'occult' (i.e. hidden) causes of phenomena. Mechanical philosophers, on the other hand, did not hesitate to invent invisible mechanisms of unseen particles to explain such phenomena as gravitation and magnetism. They believed that they were giving more plausible explanations than the scholastics, not because they had made the causes more visible but because they had made them more intelligible and, in a few important cases such as atmospheric pressure, more directly testable. Mechanical explanations, they thought, are automatically more intelligible than explanations in terms of forms and qualities: what could be easier to understand than the workings of a clock or an automaton? When Boyle invoked such analogies, as he frequently did in this treatise, he was appealing to the intuitions and experiences of an increasingly technical culture.

Nor did Boyle and many other mechanical philosophers – though Descartes would be an exception – oppose a teleological reading of nature. Indeed, Boyle vigorously defended the argument from design. But he insisted that knowledge of the first cause could never substitute for knowledge of secondary causes, which it was the primary goal of mechanistic science to attain. The main problem with the 'vulgar' notion of nature, on this score, was that its reliance on teleological explanations short-circuited the search for mechanistic ones.

Boyle seems originally to have opposed Aristotelian ideas because he perceived the naturalistic account of the world that they gave as providing the basis for irreligion. As he became committed to the mechanical philosophy during the 1650s, he laid increasing stress on the inferior intelligibility of Aristotelianism compared with the mechanical philosophy. Particularly in his *Origin of Forms and Qualities* (1666), he argued that 'form', 'substance', 'accident' and similar technical terms that were central to Aristotelianism had no real explanatory value, because they failed to explain in detail how things happened. As he put it, these made it 'very easy to solve all the phenomena of nature in general but ... impossible to explicate almost any of them in particular'.[4]

He was further hostile to Aristotelianism because it seemed to make nature purposive in an inappropriate way. To Boyle's mind, the implication of Aristotle's presumption that all bodies had a natural state, to which they inexorably returned, was that matter could think. Much of Boyle's effort in his *Free Enquiry* was devoted to proving that this was implausible. The most famous example of this is his statement that matter is incapable of obeying 'laws', in the sense of knowing whether it obeys or not; a natural 'law' is therefore 'but a notional thing', a useful explanatory tool but not a literal statement about how nature works.

For instance, it was an Aristotelian axiom that 'nature abhors a vacuum' – that nature actively worked to prevent a vacuum coming about – whereas Boyle was able to argue, deploying evidence derived from his earlier empirical investigations using his air pump, that it was in fact perfectly possible to create a vacuum, and that this was limited by purely mechanical factors. Thus when a barometer is made by filling

---

[4] *Works*, vol. 3, p. 13.

a glass tube with mercury, inverting it and submerging the open end in a pool of mercury, the liquid in the tube drops until the pressure of mercury within it balances the pressure of the air outside the tube. Under normal atmospheric conditions, the mercury stands about seventy-six centimetres (thirty inches) high. The space in the tube above the mercury appears to be empty of all substances, thus constituting a vacuum – although this was contested at the time by those who thought that a subtle fluid or an immaterial spirit might fill that space. In Boyle's opinion, phenomena such as this demonstrated not only that a vacuum can be created, but that the strength of the vacuum – the height of the column of mercury in the tube – was determined only by the pressure of the atmosphere, not by the ability of 'nature' to prevent it.

Though we have dwelt at length here on Aristotelianism, as the intellectual tradition which caused Boyle most concern, it is also important to note his hostility to Galenism, the system of ideas codified by the Greco-Roman physician, Galen (*c.* 130–200), which applied ideas similar to those of Aristotle to the medical realm. Again, in its presumption that the human body had a balance of humours which represented a healthy state, with imbalance resulting in sickness, it seemed to give matter a mind of its own. Boyle particularly took issue with the medical doctrine associated with the father of Greek medicine, Hippocrates, on whose ideas Galen had built, that 'nature is the curer of diseases': he considered this questionable on the grounds that physicians were in fact quite selective as to what they accepted of the responses of 'nature' to bodily malfunction, and sometimes felt obliged to resist these. He was similarly hostile to the related doctrine of 'critical days' in acute diseases, turning-points which took the form of the sudden excretion of harmful humours which were supposed to mark a decisive shift by nature either towards recovery or towards death. He illustrated how experience suggested that, in fact, 'nature' was not as wise as such views implied.

Boyle perceived similar dangers in certain alternative philosophies of nature which flourished in the early modern period. One was a form of Platonism, inspired by the revival of Plato's ideas that had begun in Renaissance Italy, but exemplified close to home by the group known as the Cambridge Platonists, and especially its leading lights, Henry More (1614–87) and Ralph Cudworth (1617–88), both of them known to

Boyle. They proposed a type of atomism that went beyond the standard dualism of Descartes, Boyle and others, which distinguished 'stupid', inert matter from conscious minds. Instead, they postulated what Cudworth called the 'general plastic nature of the universe' and various 'particular plastic powers in the souls of animals' – intermediaries between God and the world whose operation produced all activity, both organic and inorganic. Ultimately, this was their way of dealing with the classic theological problem of explaining how God is to be set apart from nature while maintaining that God acts within nature. More and Cudworth refused to accept either that everything happens purely by chance, or that God does everything immediately and miraculously; instead, for them plastic nature constituted God's 'inferior and subordinate instrument' that 'doth drudgingly execute that part of his providence which consists in the regular and orderly motion of matter'.[5] This was just the kind of reification of nature to which Boyle was opposed.

Another alternative that worried Boyle was the idea of active, thinking matter which was prominent in such medical authors as William Harvey (1578–1657) and Francis Glisson (1597-1677), whose writings were heavily influenced by the immanentist teleology of Aristotle and Galen. Harvey believed that vegetative activity precedes the appearance of the soul, so matter itself must possess a primitive ability to perceive, in order to organise itself. His disciple Glisson used the concept of the 'irritability' of tissues to explain how muscle fibres contract (such as how the heart contracts in response to its stretching in diastole) and how glands secrete fluids (such as how the gall-bladder releases bile when it is 'irritated' by overfilling and how the stomach digests food). He provided a metaphysical basis for this attribute in his *Treatise on the Energetic Nature of Substance* (1672), which – though little known – constitutes one of the most original philosophical systems to be devised in the second half of the seventeenth century. Glisson made vitality an essential attribute of all matter, whether living or non-living; life was just 'the energetic nature of substance' diffused throughout the universe. All things are made of a living substance capable of perception as well as motion; it shaped the parts of plants and animals, and it was the basis of nature's ability to heal itself.

---

[5] Ralph Cudworth, *The True Intellectual System of the Universe* (London, 1678), pp. 150, 171–2.

In a milder way, Boyle was also concerned about comparable implications in the world view of the Flemish chemical philosopher, J. B. van Helmont (1579–1644), which was highly influential at the time and to which Boyle himself owed a good deal. His concern was particularly about the extent to which van Helmont's philosophy – which contradicted Aristotle by arguing that form was intrinsic to matter and that it was possible to penetrate to the seminal essence, or *entelecheia*, of an object – had the implication of making nature a self-sufficient entity.

Only the mechanical philosophy, in Boyle's view, banished all chimeras of an intelligent nature, and this seemed to him a powerful argument in its favour. In his opinion, it had two main advantages. First, by giving a more coherent and intelligible explanation of natural phenomena, it held out the possibility of genuine progress in the medical and mechanical arts, potentially fulfilling the utopian vision of Bacon that was never far from Boyle's thoughts. Second, by removing intermediaries between God and the world, the mechanical philosophy benefited theology by underscoring divine sovereignty and fighting against the paganistic tendencies of Neoplatonism and related ideas.

Boyle's *Free Enquiry*

It was to assert this that Boyle wrote the treatise here presented. The theme to which it gives expression was first enunciated in a section of his *Usefulness of Natural Philosophy*, written in the 1650s and published in 1663. As far as the *Free Enquiry* itself is concerned, Boyle states in his preface that he initially wrote a substantial portion of the work in the mid-1660s, and this is confirmed by the fact that manuscript drafts survive of one section of the text written by amanuenses whom Boyle is known to have employed at that time. The work is perhaps to be seen as a kind of sequel to *The Origin of Forms and Qualities*, written a few years before it was published in 1666, which expounded Boyle's reservations about various aspects of Aristotelian doctrine, and to which the *Free Enquiry* is conceptually related. However, Boyle then put the work to one side (like other theological and philosophical works written at that time, such as his *The Excellency of Theology, compared with Natural Philosophy*, written in 1665 but published only in 1674). Only around 1680 did he return to *A Free Enquiry*, adding various passages to it, partly to clarify and reiterate its original themes, and partly to deal with threats that seemed to him newly significant.

Manuscript drafts survive of a number of the passages that Boyle added to the book at this time, and of these a disproportionate number deal with arguments of medical origin for the purposiveness of nature. This suggests that, between his initial bout of work on the book in the 1660s and this second one, Boyle had become increasingly concerned about the reification of nature in a medical context. It is almost certainly not coincidental that Glisson's *Treatise on the Energetic Nature of Substance*, which expressed such views, had appeared during the intervening years. But the passages in question appear also to reflect Boyle's concern about the commonplace presumptions about nature's role to be found among doctors at the time, reflecting the active interest in medical matters on Boyle's part which has already been referred to.

The preface to the work is dated 29 September 1682, and it seems likely that the bulk of Boyle's revision to the work had been carried out by that time. But he appears to have continued to tinker with the book right up to the last minute – literally as the copy was being set in type. We know this because the draft for a section of text that was inserted in the latter part of the book survives on the back of an Advertisement that was included in the published edition, apologising for an oversight that had occurred as the book was being printed.[6] The error to which the Advertisement drew attention was that two sections of text were printed out of order, a hardly surprising error in view of Boyle's method of composing the text in separate sections. In this edition, we have silently corrected the order of the text as instructed in this Advertisement and have indicated this in our notes.[7]

The actual publication of the book occurred early in the summer of 1686, three and a half years after the date of the preface, and the reasons for this delay are unclear. A Latin edition, advertised later that year and published in 1687, was executed by a former Jesuit who had become a Protestant and a protégé of Boyle's, the Scottish physician David Abercromby (d. 1702).[8] It provides further evidence of revisions made during the latter stages of the book's preparation, since various passages

---

[6] MS 190, fol. 6.   [7] See below, pp. xxxiv, 68, 71–3, 78, 103–4.

[8] On Abercromby, see Edward B. Davis, 'The Anonymous Works of Robert Boyle and the *Reasons Why a Protestant Should not Turn Papist* (1687)', *Journal of the History of Ideas*, 55 (1994), 611–29.

which appear in the English edition – ranging from a couple of words to whole paragraphs – fail to appear here, evidently because they were added to the text after Abercromby was given his copy to translate.[9] The Latin edition also differs from the English in being furnished with a curious adulatory poem to Boyle by 'S.F., an English noblewoman', and a fulsome panegyric of the work – a kind of publisher's blurb – by the translator. In it, Abercromby concluded: 'I therefore look upon this work as the new system of a new philosophy which fundamentally overthrows the foundation – namely, nature – of all views hitherto held in philosophical matters.'[10] The Latin edition was reissued in 1688, and it was issued twice by the Genevan publisher, de Tournes (who produced various Latin editions of Boyle's works) in the same year. The work was not reprinted in English prior to its inclusion in collected editions of Boyle's works in the eighteenth century.

The book's impact and legacy

An adulatory review of *Free Enquiry* appeared in no. 181 of *Philosophical Transactions*, for 25 May 1686; as usual with Boyle's books, this acted as a kind of promotional puff for it, being mainly devoted to a summary of the work's content.[11] A copy of the work was produced at a meeting of the Royal Society a week later, on 2 June. The task of reporting on it was given to the Cambridge don, Nathaniel Vincent (d. 1722), who duly brought an account of it to the following meeting, the content of which is not recorded.[12]

Of the attention that we know the work attracted, at least some was critical, and it is worth stressing that – although it is easy in retrospect to see the work as giving classic expression to the predominant view of nature associated with seventeenth-century science – Boyle may well have been right in seeing it as bold and paradoxical, and in being apprehensive as to how it was likely to be received. In the year of publication, for instance, Boyle received a letter from the Durham virtuoso, Peter Nelson, 'proposing some objections ... about his treatise on the vulgar notion of nature', though unfortunately this has not

---

[9] Two sentences (only) appear in the Latin edition but not in the English: see pp. 62, 165.
[10] D[avid] A[bercromby], in Lat., sig. A4v. For the poem by 'S.F.', see sigs. A6–7.
[11] Vol. 16, pp. 116–18.
[12] Thomas Birch, *The History of the Royal Society of London* (4 vols., London, 1756–7), vol. 4, pp. 487, 490.

survived.[13] What does survive is a document apparently comprising the critical remarks on it of an anonymous French Aristotelian, who took exception to several things, especially the way in which Boyle interpreted scholastic axioms in section IV. In his opinion, 'well known and well used phrases are chosen which, twisted to a new meaning, seem to allow a certain scope for error'.[14]

The most public response to the work came from a leading member of the school of thought which was among Boyle's targets, namely the Cambridge Platonists. This was the naturalist and divine, John Ray (1627–1705), whose reference to Boyle's book appeared in his *Wisdom of God Manifested in the Works of the Creation*, one of the most popular works of natural theology of the period, which was first published in 1691 and went into ten editions between then and 1735. Ray quoted Boyle's 'hypothesis' from the *Free Enquiry*, and went on to express his disagreement with it, preferring instead the idea of a 'plastic nature'. This passage recurred in all editions of Ray's work, notwithstanding his addition to the fourth edition (1704) of a further passage acknowledging how he had learned from reading Boyle's *The Christian Virtuoso* the extent of their commitment to shared theistic goals.[15]

On the other hand, the stock of Boyle's book may have risen with that of the mechanical philosophy in the aftermath of the publication of Newton's *Principia* (1687), and the subsequent popularisation of Newton's ideas. In Ephraim Chambers's *Cyclopaedia* (1728 and subsequent editions), for instance, the entry under 'nature' makes extensive reference to Boyle's 'precise treatise of the *vulgarly received Notion of Nature*'.[16] But it is perhaps symptomatic that the entry ends by summarising the three 'laws of nature' as adumbrated by Newton: this is typical of the way in which Boyle's work was conflated with Newton's in the eighteenth century, as has been observed (for instance) in his impact on the Scottish Enlightenment and hence on Hume.[17]

---

[13] Quoted from BP 36, fol. 145, a list of Boyle letters made in the early eighteenth century by Henry Miles, many of which are now lost; he described it as 'a Curious Philosoph[ical] Letter'.

[14] BP 44, fol. 54; fols. 51–4 are a set of Latin notes entitled 'Animadversions on the Aristotelian Definition of Nature'.

[15] 1st edn. (London, 1691), pp. 32–3; 4th edn. (London, 1704), pp. 53–7.

[16] *Cyclopaedia: or, an Universal Dictionary of Arts and Sciences* (2 vols., London, 1728), vol. 2, pp. 617–18.

[17] Michael Barfoot, 'Hume and the Culture of Science in the Early Eighteenth Century', in M. A. Steward (ed.), *Studies in the Philosophy of the Scottish Enlightenment* (Oxford, Clarendon Press, 1990), pp. 151–90, esp. p. 162.

The work's impact in Europe was similarly mixed. Boyle sent a copy to the German *savant*, Johann Christoph Sturm (1635–1703), who, in acknowledging it, told Boyle that he considered it 'a golden thing'.[18] Sturm was subsequently to adduce Boylian ideas in his own *De naturae agentis idolo* (1692; 'On the illusion that nature is an agent') and his *De natura sibi incassum vindicata* (1698; 'On nature employed to no purpose'), with which issue was taken by the Aristotelian, G. C. Schelhammer (1649–1716), who published *Natura sibi et medicis vindicata* (1697; 'Nature and its value to doctors') and *Naturae vindicatae vindicatio* (1702; 'A vindication of the resort to nature'). It was into this debate that G. W. Leibniz (1646–1716) intervened with his *De Ipsa Natura* ('On nature itself'), originally published in the *Acta Eruditorum* of September 1698, in which he defined the way in which he himself saw being as inherent in nature.[19]

In the twentieth century, Boyle's book has mainly been the preserve of historians of science, who have expounded his hostility to the prevalent reification of nature and – particularly since the early 1970s – have used the work to illustrate the degree to which Boyle's views were influenced by his voluntaristic theological orientation. However, the *Free Enquiry* may yet prove to be of more than merely historical significance, for some of the issues it addresses have recently come back into prominence as the mechanical view of nature that Boyle helped to establish has come under fire. When certain feminists and environmentalists advocate abandoning the modern concept of nature as an impersonal thing in favour of 'Mother Nature', they sound very much like Boyle's adversaries. When certain theologians think of God not as a transcendent creator but as the soul of the world, they echo the pagan philosophers of antiquity and the Cambridge Platonists of Boyle's day, both of whom he opposed. Proponents and opponents of these contemporary positions would do well to consider the arguments of this subtle book.

---

18 Sturm to Boyle, 1 March 1688, Royal Society Boyle Letters 5, fol. 137.
19 See Catherine Wilson, 'De Ipsa Natura. Sources of Leibniz' Doctrines of Force, Activity and Natural Law', *Studia Leibnitiana*, 19 (1987), 148–72, and Guido Giglioni, 'Automata Compared: Boyle, Leibniz and the Debate on the Notion of Life and Mind', *British Journal for the History of Philosophy*, 3 (1995), 249–78.

Chronology

1668	Settles in London, living for the rest of his life in Pall Mall, where he shared a house with his sister, Katherine, Viscountess Ranelagh
1670	Suffers severe stroke
1674	*The Excellency of Theology, compared with Natural Philosophy* published (written 1665)
1682	Dedication to *A Free Enquiry*, 29 September
1686	*A Free Enquiry into the Vulgarly Received Notion of Nature* published
1688	*A Disquisition about the Final Causes of Natural Things* published
1690	*The Christian Virtuoso* published
1691	Boyle dies, 31 December
1692	Boyle's funeral at St Martins-in-the Fields, 7 January

Further reading

Our understanding of Boyle's thought has in recent years undergone a major transformation. For an essay-collection which indicates the kind of work currently in progress on different aspects of Boyle's ideas, with an introduction which attempts to draw some overall conclusions about his intellectual personality, see Michael Hunter (ed.), *Robert Boyle Reconsidered* (Cambridge, Cambridge University Press, 1994). One of the essays in this has since been developed into a useful book-length account of Boyle's scientific outlook, Rose-Mary Sargent, *The Diffident Naturalist: Robert Boyle and the Philosophy of Experiment* (Chicago, University of Chicago Press, 1995), though she underestimates the significance of the voluntarist streak in his thought which is so important to a proper understanding of his *Free Enquiry*.

This oversight can be rectified by attention to J. E. McGuire's key study, 'Boyle's Conception of Nature', *Journal of the History of Ideas*, 33 (1972), 523–42, which first emphasised this facet of Boyle's ideas. Also interesting in this connection is Margaret Osler, 'The Intellectual Sources of Boyle's Philosophy of Nature: Gassendi's Voluntarism and Boyle's Physico-theological Project', in Richard Kroll, Richard Ashcraft and Perez Zagorin (eds.), *Philosophy, Science and Religion in England 1640–1700* (Cambridge University Press, 1992), pp. 178–98. A further study which valuably places Boyle's views on God's relations with nature in the context of those of his contemporaries is Keith Hutchison, 'Supernaturalism and the Mechanical Philosophy', *History of Science*, 21 (1983), 297–333. See also Richard Olson, 'On the Nature of God's Existence, Wisdom and Power: The Interplay between Organic and Mechanistic Imagery in Anglican Natural Theology, 1640–1740', in

Frederick Burwick (ed.), *Approaches to Organic Form* (Dordrecht, Reidel, 1987), pp. 1–48.

For fuller information on Boyle's life and intellectual evolution, see two works by Michael Hunter: 'How Boyle Became a Scientist', *History of Science*, 33 (1995), 59–103 (which deals with the transformation in his interests that occurred around 1650) and *Robert Boyle by Himself and his Friends* (London, Pickering & Chatto, 1994), an edition of the key contemporary memoirs of Boyle, including his own autobiographical 'Account of Philaretus during his Minority'. On Boyle's earlier, literary phase, see Lawrence Principe, 'Virtuous Romance and Romantic Virtuoso: The Shaping of Robert Boyle's Literary Style', *Journal of the History of Ideas*, 56 (1995), 377–97, while for the texts of some of the works that Boyle wrote at that time, see John T. Harwood (ed.), *The Early Essays and Ethics of Robert Boyle* (Carbondale and Edwardsville, Southern Illinois University Press, 1991).

On English science in Boyle's period, a useful synthesis is John Henry, 'The Scientific Revolution in England', in Roy Porter and Mikulas Teich (eds.), *The Scientific Revolution in National Context* (Cambridge University Press, 1992), pp. 178–209. On the polemical preoccupations of scientists and their anxiety about the threat of 'atheism', see Michael Hunter, 'Science and Heterodoxy: An Early Modern Problem Reconsidered', in D. C. Lindberg and R. S. Westman (eds.), *Reappraisals of the Scientific Revolution* (Cambridge University Press, 1990), pp. 437–60, reprinted in Hunter, *Science and the Shape of Orthodoxy: Intellectual Change in late Seventeenth-century Britain* (Woodbridge, Boydell Press, 1995), pp. 225–44. Further information on the context and organisation of the new science will be found in Michael Hunter, *Science and Society in Restoration England* (Cambridge University Press, 1981), which also includes a bibliographical essay surveying the secondary literature up to 1981; this is updated by the foreword to the 1992 reprint of the book (Aldershot, Gregg Revivals, 1992), pp. viii–xvi. An influential, more recent work is Steven Shapin and Simon Schaffer, *Leviathan and the Air Pump: Hobbes, Boyle and the Experimental Life* (Princeton, Princeton University Press, 1985), which uses Boyle's controversy with Thomas Hobbes to reconsider issues concerning the production of knowledge and the validation of knowledge claims in the period.

Various studies exist of the alternative philosophies of nature which flourished in Boyle's day and with which he took issue in *A Free Enquiry*. On the Aristotelian tradition a useful recent synthesis is provided by Christia Mercer, 'The Vitality and Importance of Early Modern Aristotelianism', in Tom Sorell (ed.), *The Rise of Modern Philosophy* (Oxford, Clarendon Press, 1993), pp. 33–67. For a further brief overview, see A. G. Molland, 'Aristotelian Science', in R. C. Olby, G. N. Cantor, J. R. R. Christie and M. J. S. Hodge (eds.), *Companion to the History of Modern Science* (London, Routledge, 1990), pp. 557–67.

On the neo-Aristotelian cosmologies of the 1670s which caused Boyle concern, see Alan Cromartie, *Sir Matthew Hale 1609–76: Law, Religion and Natural Philosophy* (Cambridge University Press, 1995); John Henry, 'Medicine and Pneumatology: Henry More, Richard Baxter and Francis Glisson's *Treatise on the Energetic Nature of Substance*', *Medical History*, 31 (1987), 15–40; and Guido Giglioni, 'The "Hylozoic" Foundations of Francis Glisson's Anatomical Research', in O. P. Grell and Andrew Cunningham (eds.), *Religio Medici: Medicine and Religion in Seventeenth-century England* (Aldershot, Scolar Press, forthcoming).

On Boyle's disagreement with the Cambridge Platonist, Henry More, see especially John Henry, 'Henry More versus Robert Boyle: The Spirit of Nature and the Nature of Providence', in Sarah Hutton (ed.), *Henry More (1614–87): Tercentenary Studies* (Dordrecht, Kluwer Academic Publishers, 1990), pp. 55–76. For a useful account of the thought of Ralph Cudworth in this connection, see the articles by Hutchison and Olson already cited. An older study which deals with these and other thinkers and retains some value is E. A. Burtt, *The Metaphysical Foundations of Modern Physical Science* (rev. edn., London, Kegan Paul, 1932).

For the claim that *A Free Enquiry* was aimed at a putative group of 'pagan naturalists' led by Henry Stubbe, see J. R. Jacob, 'Boyle's Atomism and the Restoration Assault on Pagan Naturalism', *Social Studies of Science*, 8 (1978), 211–33, largely repeated in his *Henry Stubbe, Radical Protestantism and the Early Enlightenment* (Cambridge University Press, 1983). For a critique of these claims see Nicholas H. Steneck, 'Greatrakes the Stroker: The Interpretations of Historians', *Isis*, 73 (1982), 161–77, many of whose reservations are well founded notwithstanding the response to them by Jacob in the postscript to *Henry Stubbe*.

Lastly, for a lengthy analysis of the draft material relating to *A Free*

Enquiry that survives among the Boyle Papers and its implications in terms both of how the book took shape and of Boyle's method of composition more generally, see Michael Hunter and Edward B. Davis, 'The Making of Robert Boyle's *Free Enquiry into the Vulgarly Receiv'd Notion of Nature* (1686)', *Early Science and Medicine*, 1 (1996), 204–71. This also reassesses the claims of Jacob concerning the identity of the 'sect of men' noted above, in the course of surveying the various threats that preoccupied Boyle at different stages in the work's composition.

A note on the text

A modernised text of sections II and IV of *Free Enquiry* (except for the excursus on ancient polytheism in the latter) is to be found in M. A. Stewart (ed.), *Selected Philosophical Papers of Robert Boyle* (Manchester, Manchester University Press, 1979; reprint Indianapolis, Hackett Publishing Co., 1991), pp. 176–91. For the complete work, the most accessible text has been in Thomas Birch's eighteenth-century edition of Boyle's *Works*: the work appears in vol. 4 of the first edition (1744), and vol. 5 in the second edition (1772; reprint Hildesheim, Georg Olms, 1965). Birch's text is modernised to eighteenth-century taste; in addition, Birch failed to find the Advertisement referred to above (which is missing in many copies), and he therefore did not follow Boyle's instructions for reordering certain passages (see above, p. xxiv).

This text is based on the first edition (1686), but Boyle's instructions have been followed in reinstating the misplaced sections where they belong. We have also made an editorial change which both Boyle and Birch failed to effect, and this is to delete obsolete section numbers which appeared at the start of the first line of text in sections II to VI both in the first edition and in Birch's edition. These numbers run one behind those actually allocated to the sections in question; they should have been superseded when Boyle decided to give the number 'I' to what must originally have been an unnumbered introductory section, and their retention would be liable to confuse the reader.

We have collated the text with the 1687 Latin edition, noting variants between the two in the abbreviated form 'Lat. has' or 'Lat. lacks'. We have also followed variants in extant manuscript drafts where they make better sense than the published text, calling attention to these. Such matters are dealt with in editorial footnotes, which are assigned letters.

Boyle's own notes are reproduced with minor editorial changes and assigned arabic numerals. His references are often somewhat obscure, and we have provided additional information in square brackets where appropriate. Occasionally, his quotations from classical and other works represent an adaptation of the original, and we have quoted the original Latin, in such cases giving an English translation in square brackets.

Following the editorial policy of the Cambridge Texts in the History of Philosophy, Boyle's text has been modernised throughout, according to the following conventions:

Capitalisation and spelling have been consistently modernised, including the removal of archaic contractions such as 'tis'. The ampersand in Latin has been replaced by 'et'.

Italics have been removed except for quotations from Latin books, words in languages other than English and rare instances in which the use of italics for emphasis adds significantly to the clarity. For passages in English which Boyle intended to present as quotations, inverted commas have been substituted.

Sentences have been repunctuated where necessary to give a better flow and sense. This has involved the frequent deletion, and occasional insertion, of commas. Elsewhere, we have altered commas to colons or semicolons, and have sometimes placed phrases within dashes (which Boyle did not use) for greater clarity. Where desirable and possible without inserting or omitting words, we have broken up excessively long sentences, especially by substituting full stops for Boyle's own semicolons and colons.

We have altered the location of paragraph breaks where it improves readability.

Square brackets have been used to indicate words or translations inserted by the editors. On the rare occasions where square brackets are Boyle's own, this is indicated in the notes.

Boyle's usage of 'nature' has presented us with something of a problem, since in the first edition this word is repeatedly placed in italics and capitalised. For the sake of consistency, we have refrained throughout from italicising or capitalising the word. Boyle also often uses *Free*

Enquiry or *Enquiry* in italics to refer to the book; in this case, we have retained his italics.

An unmodernised text of the work will be presented in the forthcoming 'Pickering Masters' edition of *The Works of Robert Boyle*, ed. Michael Hunter and Edward B. Davis (London: Pickering & Chatto). This will also include transcripts of material which Boyle drafted for the work *c.* 1680 – including a disquisition on desirable and undesirable forms of atomism – but which he failed to include in it, probably because he considered it tangential to the work's main theme.

A glossary of unfamiliar terms used by Boyle is found on pp. 166–7.

A
FREE ENQUIRY
Into the Vulgarly Received
NOTION
OF
NATURE
Made in an
ESSAY
Addressed to a FRIEND

By R[obert] B[oyle] Fellow of the Royal Society

Audendum est, & veritas investiganda;
quam etiamsi non assequamur, omnino tamen
propius, quam nunc sumus, ad eam perveniemus.
Galenus

[One must be daring and approach the truth:
for even if we may not grasp it completely,
yet we will get closer to it than we are now.
Galen][a]

---

^a Boyle used the same quotation from Galen on the title-page of *The Origin of Forms and Qualities* (1666). We have not located the exact place in Galen's voluminous writings from which it is derived.

The Preface

I have often wondered that, in so inquisitive an age as this, among those many learned men that have with much freedom, as well as acuteness, written of the works of nature (as they call them) – and some of them of the principles too – I have not met with any that has made it his business to write of nature herself. This will perhaps hereafter be thought such an omission as if, in giving an account of the political estate of a kingdom, one should treat largely of the civil judges, military officers and other subordinate magistrates, and of the particular ranks and orders of inferior subjects and plebeians, but should be silent of the prerogatives and ways of administration of the king; or (to use a comparison more suitable to the subject) as if one should particularly treat of the barrel, wheels, string, balance, index and other parts of a watch, without examining the nature of the spring that sets all these a-moving. When I say this, I do not forget that the word 'nature' is everywhere to be met with in the writings of physiologers. But though they frequently employ the word, they seem not to have much considered what notion ought to be framed of the thing, which they suppose and admire, and upon occasion celebrate, but do not call in question or discuss.

Weighing therefore with myself of what great moment the framing a right or a wrong idea of nature must be in reference to both the speculative and practical part of physiology, I judged it very well worth the while to make, with philosophical freedom, a serious *Enquiry* into the vulgarly received notion of nature – that, if it appeared well grounded, I might have the rational satisfaction of not having acquiesced in it till after a previous examen; if I should find it confused and ambiguous, I might endeavour to remedy that inconvenience by distinguishing the acceptions of the word; if I found it dubious as to its truth, I might be shy in trusting too much to a distrusted principle; and, if I found [it] erroneous, I might avoid the raising superstructures of my own, or relying on those of others, that must owe their stability to an unsound and deceitful foundation.

And because many atheists ascribe so much to nature that they think it needless to have recourse to a deity for the giving an account of the phenomena of the universe: and on the other side, very many theists[a] seem to think the commonly received notion of nature little less than necessary to the proof of the existence and providence of God, I – who differ from both these parties and yet think every true theist,[b] and much more every true Christian, ought to be much concerned for truths that have so powerful an influence on religion – thought myself, for its sake, obliged to consider this matter, both with the more attention and with regard to religion.

And yet, being to write this treatise as a physiologer, not a Christian, I could not rationally build any positive doctrine upon mere revelation, which would have been judged a foreign principle in this *Enquiry*. Only, since the person I intentionally addressed my thoughts to, under the name of Eleutherius,[c] was a good Christian, I held it not impertinent now and then, upon the by, to intimate something to prevent or remove some scruples that I thought he might have on the score (I say not of natural theology, for that is almost directly pertinent, but)[d] of the Christian faith. But these passages are very few, and but transiently touched upon.

Since the reader will be told by and by both that and why the papers that make up the following treatise were not written in one continued series of times, but many years were interposed between the writing of some of them and that of those which precede and follow them, I hope it will be thought but a venial fault if the contexture of the whole discourse do not appear so uniform, nor all the connections of its parts so apt and close, as (if no papers had been lost and supplied) might reasonably be looked for.

I expect the novelty of divers of the sentiments and reasonings proposed in the following discourse will be surprising, and incline many to look upon the author as a bold man, and much addicted to paradoxes. But, having formerly in a distinct essay delivered my thoughts about paradoxes in general,[e] I shall not now engage in that subject, but confine myself to what concerns the ensuing paper. I say then in short that, in an opinion, I look upon its being new or ancient, and its being singular or commonly received, as things that are but extrinsical to its being true or false. And, as I would never reject a truth for being generally known

---

[a] Here Lat. adds: 'or not a few Deists' (*seu Deistae non pauci*). [b] Lat. has: 'Deist'.

[c] Lat. lacks 'under the name of Eleutherius'. The identity of Eleutherius has not been established.

[d] Lat. lacks the parenthetical phrase.

[e] This work was not published and has not survived.

or received, so will I not conclude an opinion to be a truth merely because great numbers have thought it to be so, nor think an opinion erroneous because it is not yet known to many or because it opposes a tenent embraced by many. For I am wont to judge of opinions as of coins: I consider much less, in any one that I am to receive, whose inscription it bears, than what metal it is made of. It is indifferent enough to me whether it was stamped many years or ages since, or came but yesterday from the mint. Nor do I regard through how many, or how few, hands it has passed for current, provided I know by the touchstone or any sure trial purposely made, whether or no it be genuine, and does or does not deserve to have been current. For if upon due proof it appears to be good, its having been long and by many received for such will not tempt me to refuse it. But if I find it counterfeit, neither the prince's image or inscription, nor its date (how ancient soever), nor the multitude of hands through which it has passed unsuspected will engage me to receive it. And one disfavouring trial, well made, will much more discredit it with me than all those specious things I have named can recommend it.

By this declaration of my sentiments about paradoxes in general, I hope it will be thought that the motive I had to question that notion of nature which I dissent from, was not that this notion is vulgarly received. And I have this to say, to make it probable that I was not engaged in this controversy by any ambition of appearing in print an heresiarch in philosophy, by being the author of a strange doctrine: that the following discourse was written about the year 1666 (that is, some lustres ago); and that not long after, the youth to whom I dictated it having been inveigled to steal away, unknown to me or his parents, into the Indies (whence we never heard of him since), left the loose sheets, wherein (and not in a book) my thoughts had been committed to paper, very incoherent by the omission of divers necessary passages. Upon which account – and my unwillingness to take the pains to supply what was wanting – those papers lay by me many years together neglected and almost forgotten, until the curiosity of some philosophical heads that were pleased to think they deserved another fate obliged me to tack them together and make up the gaps that remained between their parts by retrieving – as well as after so many years my bad memory was able to do – the thoughts I sometimes had pertinent to those purposes. And indeed, when I considered of how vast importance it is in philosophy –

and the practice of physic too – to have a right notion of nature, and how little the authority of the generality of men ought, in so nice and intricate a subject, to sway a free and impartial spirit, as I at first thought myself obliged, since others had not saved me the labour, to make a *Free Enquiry* into this noble and difficult subject, so I was afterwards the more easily prevailed with by those that pressed the publication of it.

With what success I have made this attempt, I must leave others to judge. But if I be not much flattered, whatever becomes of the main attempt, there will be found suggested here and there in the following discourse some reflections and explications that will at least oblige the zealous assertors of the vulgar notion of nature to clear up the doctrine, and speak more distinctly and correctly about things that relate to it than hitherto has been usual. And that will be fruit enough to recompense the labour and justify the title of a *Free Enquiry*. In prosecution of which – since I have been obliged to travel in an untrodden way without a guide – it will be thought (I hope) more pardonable than strange if, in attempting to discover divers general mistakes, I be not so happy as to escape falling into some particular ones myself. And if among these, I have been so unhappy as make any that is injurious to religion, as I did not at all intend it, so, as soon as ever I shall discover it, I shall freely disown it myself and pray that it may never mislead others. What my performance has been, I have already acknowledged that I may be unfit to judge; but for my intentions, I may make bold to say, they were to keep the glory of the divine author of things from being usurped or entrenched upon by his creatures, and to make his works more thoroughly and solidly understood by the philosophical studiers of them.

I do not pretend – and I need not – that every one of the arguments I employ in the following tract is cogent, especially if considered as single. For demonstrative arguments would be unsuitable to the very title of my attempt; since if about the received notion of nature I were furnished with unanswerable reasons, my discourse ought to be styled not a *Free Enquiry* into the vulgar notion of nature I consider, but a *Confutation* of it. And a heap of bare probabilities may suffice to justify a doubt of the truth of an opinion which they cannot clearly evince to be false. And therefore, if any man shall think fit to criticise upon the less principal or less necessary parts of this treatise, perhaps I shall not think myself obliged to be concerned at it. And even if the main body of the discourse itself shall be attacked from the press, I, who am neither young nor healthy, nor ever made divinity, philosophy or physic my profession, am not like to oppose him in the same way. Since, as I ought not to wish that any errors of mine (if this essay teach any such) should prevail, so, if the things I have delivered be true for the main, I need not despair but that – in such a free and inquisitive age as ours – there will be found generous spirits that will not suffer weighty truths to be oppressed – though the proposers of them should, by averseness from contention or by want of time or health, be themselves kept from defending them. Which I have thought fit to take notice of in this place, that the truth (if I have been so happy as to have found and taught it) may not suffer by my silence; nor any reader surmise that, if I shall leave a book unanswered, I thereby acknowledge it to be unanswerable.

But this regards only the main substance of our essay, not the order or disposition of the parts – since if any shall censure that, I shall not quarrel with him about it. For indeed, considering in how preposterous an order the papers I have here tacked together came to hand, and how many things are upon that score unduly placed, I shall not only be content, but must desire, to have this rhapsody of my own loose papers looked upon but as an apparatus, or collection of materials, in order to [what I well know this maimed and confused essay is not[f]] a complete and regular discourse. Yet, to conclude, I thought that the affording even of a little light, in a subject so dark and so very important, might keep an essay from being useless, and that to fall short of demonstration would prove a pardonable fault in a discourse that pretends not to dogmatise but only to make an *Enquiry*.

September 29, 1682[g]

---

[f]   Lat. lacks this phrase; the square brackets are in the original.
[g]   The 'Advertisement' discussed in the Introduction was intended to go here, as a catchword at
      the end of the Preface shows.

A
Free Enquiry
Into the Received
NOTION
OF
NATURE

SECTION I

I know not whether or not it be a prerogative in the human soul, that, as
it is itself a true and positive being, so it is apt to conceive all other
things as true and positive beings also. But whether or no this
propensity to frame such kind of ideas suppose an excellency, I fear it
occasions mistakes, and makes us think and speak after the manner of
true and positive beings of such things as are but chimerical, and some
of them negations or privations themselves, as death, ignorance, blind-
ness and the like. It concerns us therefore to stand very carefully upon
our guard, that we be not insensibly misled by such an innate and
unheeded temptation to error as we bring into the world with us. And
consequently I may be allowed to consider whether – among other
particulars in which this deluding propensity of our minds has too great
(though unsuspected) an influence upon us – it may not have imposed
on us in the notion we are wont to frame concerning nature. For this
being the fruitful parent of other notions, as nature herself is said to be
of the creatures of the universe, the notion is so general in its
applications, and so important in its influence, that we had need be
jealously careful of not over easily admitting a notion, than which there
can scarce be any that more deserves to be warily examined before it be
thoroughly entertained.

Let me therefore make bold to enquire freely, whether that of which
we affirm such great things, and to which we ascribe so many feats, be
that almost divine thing whose works among others we are, or a notional
thing that in some sense is rather to be reckoned among our works as
owing its being to human intellects. I know most men will be forestalled
with no mean prejudices against so venturous an attempt, but I will not

do Eleutherius the injury to measure him by the prepossessed generality of men. Yet there are two scruples which I think it not amiss to take notice of, to clear the way for what shall be presented you in the following discourse.

And first, it may seem an ungrateful and unfilial thing to dispute against nature, that is taken by mankind for the common parent of us all. But though it be an undutiful thing to express a want of respect for an acknowledged parent, yet I know not why it may not be allowable to question one that a man looks upon but as a pretended one, or at least does upon probable grounds doubt whether she be so or no. And until it appear to me that she is so, I think it my duty to pay my gratitude, not to I know not what, but to that deity whose wisdom and goodness not only designed to make me a man and enjoy what I am here blessed with, but contrived the world so that even those creatures of his, who by their inanimate condition are not capable of intending to gratify me, should be as serviceable and useful to me as they would be, if they could and did design the being so. And you may be pleased to remember that, as men now accuse such an enquirer as I am of impiety and ingratitude towards nature, so the Persians and other worshippers of the celestial bodies accused several of the ancient philosophers and all the primitive Christians of the like crimes in reference to the sun, whose existence and whose being a benefactor to mankind was far more unquestionable than that there is such a semi-deity as men call nature. And it can be no great disparagement to me to suffer on the like account with so good company, especially when several of the considerations that justify them may also apologise for me. I might add that – it not being half so evident to me that what is called nature is my parent, as that all men are my brothers by being the 'offspring of God' (for the τοῦ γὰρ γένος ἐσμέν of Aratus is adopted by St Paul)[1] – I may justly prefer the doing of them a service, by disabusing them to the paying of her a ceremonial respect. But setting allegories aside, I have sometimes seriously doubted whether the vulgar notion of nature has not been both injurious to the glory of God, and a great impediment to the solid and useful discovery of his works.

And first, it seems to detract from the honour of the great author and governor of the world that men should ascribe most of the admirable

---

[1]   Acts 17. [Acts 17:28–9, from Paul's sermon on Mars' Hill in Athens, where he quotes the invocation to Zeus from the poem *Phenomena* by Aratus of Soli (*c.* 315–240 BC).]

things that are to be met within it not to him, but to a certain nature which themselves do not well know what to make of. It is true that many confess that this nature is a thing of his establishing and subordinate to him. But, though many confess it when they are asked whether they do or no – yet, besides that, many seldom or never lifted up their eyes to any higher cause – he that takes notice of their way of ascribing things to nature may easily discern that, whatever their words sometimes be, the agency of God is little taken notice of in their thoughts. And, however, it does not a little darken the excellency of the divine management of things that, when a strange thing is to be effected or accounted for, men so often have recourse to nature and think she must extraordinarily interpose to bring such things about; whereas it much more tends to the illustration of God's wisdom to have so framed things at first that there can seldom or never need any extraordinary interposition of his power. And, as it more recommends the skill of an engineer to contrive an elaborate engine so as that there should need nothing to reach his ends in it but the contrivance of parts devoid of understanding, than if it were necessary that ever and anon a discreet servant should be employed to concur notably to the operations of this or that part, or to hinder the engine from being out of order; so it more sets off the wisdom of God in the fabric of the universe that he can make so vast a machine perform all those many things which he designed it should by the mere contrivance of brute matter, managed by certain laws of local motion and upheld by his ordinary and general concourse, than if he employed from time to time an intelligent overseer – such as nature is fancied to be – to regulate, assist and control the motions of the parts. In confirmation of which, you may remember that the later poets justly reprehended their predecessors for want of skill in laying the plots of their plays, because they often suffered things to be reduced to that pass that they were fain to bring some deity (θεὸς ἀπὸ μηχανῆς) upon the stage to help them out. (*Nec Deus intersit, nisi dignis vindice nodus,* etc. [And let no god intervene, unless there has arisen a tangle meriting a champion.])[a]

And let me tell you freely that, though I will not say that Aristotle meant the mischief his doctrine did, yet I am apt to think that the

---

[a] The Greek is the equivalent of the Latin *deus ex machina*, i.e. a god from the machine, brought on at the end of a play to sort out a complicated plot. The Latin quotation is from Horace, *Ars Poetica*, 191.

grand enemy of God's glory[b] made great use of Aristotle's authority and errors to detract from it. For, as Aristotle, by introducing the opinion of the eternity of the world (whereof he owns himself to have been the first broacher), did (at least in almost all men's opinion) openly deny God the production of the world, so by ascribing the admirable works of God to what he calls nature, he tacitly denies him the government of the world. Which suspicion, if you judge severe, I shall not at more leisure refuse to acquaint you (in a distinct paper)[c] why I take divers of Aristotle's opinions relating to religion to be more unfriendly, not to say pernicious, to it than those of several other heathen philosophers.

And here give me leave to prevent an objection that some may make, as if, to deny the received notion of nature, a man must also deny providence, of which nature is the grand instrument. For in the first place, my opinion hinders me not at all from acknowledging God to be the author of the universe and the continual preserver and upholder of it – which is much more than the Peripatetic hypothesis, which (as we were saying) makes the world eternal, will allow its embracers to admit. And those things which the school philosophers ascribe to the agency of nature interposing according to emergencies, I ascribe to the wisdom of God in the first fabric of the universe; which he so admirably contrived that, if he but continue his ordinary and general concourse, there will be no necessity of extraordinary interpositions, which may reduce him to seem as it were to play after-games – all those exigencies, upon whose account philosophers and physicians seem to have devised what they call nature, being foreseen and provided for in the first fabric of the world; so that mere matter, so ordered, shall in such and such conjunctures of circumstances, do all that philosophers ascribe on such occasions to their almost omniscient nature, without any knowledge of what it does, or acting otherwise than according to the catholic laws of motion. And methinks the difference between their opinion of God's agency in the world, and that which I would propose, may be somewhat adumbrated by saying that they seem to imagine the world to be after the nature of a puppet, whose contrivance indeed may be very artificial,

---

[b]   i.e. the Devil.
[c]   This may be a reference to the 'Requisite Digression concerning those, that would exclude the Deity from intermeddling with matter', Essay 4 of Book I of *The Usefulness of Experimental Natural Philosophy* (Oxford, 1663), *Works*, vol. 2, pp. 36f.

but yet is such that almost every particular motion the artificer is fain (by drawing sometimes one wire or string, sometimes another) to guide, and oftentimes overrule, the actions of the engine; whereas, according to us, it is like a rare clock, such as may be that at Strasbourg,[d] where all things are so skilfully contrived that the engine being once set a-moving, all things proceed according to the artificer's first design, and the motions of the little statues that at such hours perform these or those things do not require (like those of puppets) the peculiar interposing of the artificer or any intelligent agent employed by him, but perform their functions upon particular occasions by virtue of the general and primitive contrivance of the whole engine. The modern Aristotelians and other philosophers would not be taxed as injurious to providence, though they now ascribe to the ordinary course of nature those regular motions of the planets that Aristotle and most of his followers (and among them the Christian schoolmen) did formerly ascribe to the particular guidance of intelligent and immaterial beings, which they assigned to be the movers of the celestial orbs. And when I consider how many things that seem anomalies to us do frequently enough happen in the world, I think it is more consonant to the respect we owe to divine providence to conceive that, as God is a most free as well as a most wise agent, and may in many things have ends unknown to us, he very well foresaw and thought fit that such seeming anomalies should come to pass, since he made them (as is evident in the eclipses of the sun and moon) the genuine consequences of the order he was pleased to settle in the world, by whose laws the grand agents in the universe were empowered and determined to act according to the respective natures he had given them; and the course of things was allowed to run on, though that would infer the happening of seeming anomalies and things really repugnant to the good or welfare of divers particular portions of the universe. This (I say) I think to be a notion more respectful to divine providence than to imagine, as we commonly do, that God has appointed an intelligent and powerful being called nature to be as his vicegerent, continually watchful for the good of the universe in general and of the particular bodies that compose it; while, in the meantime, this being appears not to have the skill or the power to prevent such anomalies, which oftentimes prove destructive to

---

[d] The Swiss mathematician Cunradus Dasypodius (*c.* 1530–1600) built a superb mechanical clock in the cathedral at Strasbourg between 1571 and 1574.

multitudes of animals and other noble creatures (as in plagues, etc.), and sometimes prejudicial to greater portions of the universe (as in earthquakes of a large spread, eclipses of the luminaries, great and lasting spots on the sun, eruptions of Vulcan,[c] great comets or new stars that pass from one region of heaven to another). And I am the more tender of admitting such a lieutenant to divine providence as nature is fancied to be, because I shall hereafter give you some instances in which it seems that if there were such a thing, she must be said to act too blindly and impotently to discharge well the part she is said to be trusted with.

I shall add that the doctrine I plead for does, much better than its rival, comply with what religion teaches us about the extraordinary and supernatural interpositions of divine providence. For when it pleases God to overrule or control the established course of things in the world by his own omnipotent hand, what is thus performed may be much easier discerned and acknowledged to be miraculous, by them that admit in the ordinary course of corporeal things nothing but matter and motion, whose powers men may well judge of, than by those who think there is besides a certain semi-deity which they call nature, whose skill and power they acknowledge to be exceeding great, and yet have no sure way of estimating how great they are and how far they may extend. And give me leave to take notice to you on this occasion that I observe the miracles of our Saviour and his apostles, pleaded by Christians on the behalf of their religion, to have been very differingly looked on by Epicurean and other corpuscularian infidels and by those other unbelievers who admit of a soul of the world or spirits in the stars or, in a word, think the universe to be governed by intellectual beings distinct from the supreme being we call God. For this latter sort of infidels have often admitted those matters of fact which we Christians call miracles, and yet have endeavoured to solve them by astral operations and other ways not here to be specified; whereas the Epicurean enemies of Christianity have thought themselves obliged resolutely to deny the matters of fact themselves, as well discerning that the things said to be performed exceeded the mechanical powers of matter and motion (as they were managed by those that wrought the miracles), and consequently must either be denied to have been done or be confessed to have been truly miraculous. But there

---

[c] i.e. volcanoes.

14

may hereafter be occasion[2] both to improve the things already said, and others to satisfy theological scruples about our hypothesis.

I formerly told you that it was not only to the glory of God (as that results from his wisdom, power and goodness expressed in the world) that I suspected the notion of nature that I am examining to be prejudicial, but also to the discovery of his works. And you will make no great difficulty to believe me, if you consider that, while men allow themselves so general and easy a way of rendering accounts of things that are difficult, as to attribute them to nature, shame will not reduce them to a more industrious scrutiny into the reasons of things, and curiosity itself will move them to it the more faintly – of which we have a clear and eminent example in the ascension of water in pumps and in other phenomena of that kind, whose true physical causes had never been found out if the moderns had acquiesced (as their predecessors did) in that imaginary one, that the world was governed by a watchful being called nature, and that she abhors a vacuum, and consequently is still in a readiness to do irresistibly whatever is necessary to prevent it. Nor must we expect any great progress in the discovery of the true causes of natural effects, while we are content to sit down with other than the particular and immediate ones.

It is not that I deny that there are divers things – as the number and situation of the stars, the shapes and sizes of animals, etc. – about which even a philosopher being asked can say little, but that it pleased the author of the universe to make them so. But when we give such general answers, we pretend not to give the particular physical reasons of the things proposed, but do in effect confess we do not know them. To this I add, that the veneration wherewith men are imbued for what they call nature has been a discouraging impediment to the empire of man over the inferior creatures of God. For many have not only looked upon it as an impossible thing to compass, but as something of impious to attempt: the removing of those boundaries which nature seems to have put and settled among her productions. And while they look upon her as such a venerable thing, some make a kind of scruple of conscience to endeavour so to emulate any of her works, as to excel them.

I have stayed so long about removing the first of the two scruples I

---

[2] See the third, the fourth and also the last section of this treatise. [These are sections IV, V and VIII as actually numbered in this text. The present section, though called 'I', functions as an introduction and is left out of Boyle's accounting here. See A note on the text.]

formerly proposed against my present attempt, that, not to tire your patience, I shall in few words dispatch the second – which is, that I venture to contradict the sense of the generality of mankind. To which I answer, that in philosophical inquiries it becomes not a naturalist to be so solicitious what has been, or is believed, as what ought to be so. And I have also elsewhere, on another occasion,[f] showed how little the sense of the generality of men ought to sway us in some questions. But that which I shall at present more directly reply is: first, that it is no wonder men should be generally prepossessed with such a notion of nature as I call in question, since education (especially in the schools) has imbued them with it from their infancy, and even in their maturer years they find it taken for granted and employed not only by the most, but by the learnedest writers, and never hear it called in question by any. And then it exceedingly complies with our innate propensity to think that we know more than we do, and to appear to do so. For to vouch nature for a cause is an expedient that can scarce be wanting to any man, upon any occasion, to seem to know what he can indeed render no good reason of.

And to this first part of my answer, I shall subjoin this second: that the general custom of mankind to talk of a thing as a real and positive being, and attribute great matters to it, does but little weigh with me, when I consider that – though fortune be not any physical thing but a certain loose and undetermined notion which a modern metaphysician would refer to the class of his non entities – yet not only the Gentiles made it a goddess (*Nos te facimus, Fortuna, Deam, Cæloque locamus* [We make you a goddess, Fortune, and place you in the heavens])[g] which many of them seriously worshipped, but eminent writers in verse and in prose, ethnic and Christian, ancient and modern, and all sorts of men in their common discourse do seriously talk of it as if it were a kind of Antichrist that usurped a great share in the government of the world, and ascribe little less to it than they do to nature. And not to speak of what poets, moralists and divines tell us of the powers of ignorance and vice, which are but moral defects, let us consider what things are not only by these men, but by the generality of mankind, seriously attributed to death, to which so great and fatal a dominion is assigned. And then, if we consider too that this death, which is said to do so many

---

[f] Perhaps a reference to the unpublished material on 'The Uses and Extent of Experience, Reason, and Authority in Natural Philosophy', in BP 9.

[g] Quoting (in a slightly abridged form) Juvenal, *Satires*, x.365–6.

and such wonderful things, is neither a substance nor a positive entity, but a mere privation, we shall, I trust, the less believe that the feats ascribed to nature do infer that there is really such a physical agent as is supposed.

And now, having (as I presume) cleared our enquiry as far as it is yet necessary – and it will be further done hereafter[h] – from those prejudices that might make the attempt be censured before it be examined, I proceed to the enquiry itself, wherein I shall endeavour (but with the brevity my want of leisure exacts) to do these six things. First, to give you a short account of the great ambiguity of the word 'nature' arising from its various acceptions. Secondly, to show you that the definition also that Aristotle himself gives of nature does not afford a clear or satisfactory notion of it. Thirdly, to gather from the several things that are wont to be affirmed of, or attributed to, nature, the received notion of it, which cannot be well gathered from the name because of its great ambiguity. Fourthly, I will mention some of those reasons that dissuade me from admitting this notion of nature. Fifthly, I shall endeavour to answer severally the chief things upon which men seem to have taken up the idea of nature that I disallow. And sixthly, I shall propose some of the chief *effata* or axioms that are wont to be made use of concerning nature in general, and shall show how far, and in what sense, I may admit them.

And here it may be opportune to prevent both mistakes, and the necessity of interrupting the series of our discourse, to set down two or three advertisements.

1. When anywhere in this tract I speak of the opinions of Aristotle and the Peripatetics, as I would not be thought to impute to him all the sentiments of those that will be thought his followers, some of which seem to me to have much mistaken his true meaning; so, on the other side, I did not conceive that my design obliged me to enquire anxiously into his true sentiments, whether about the origin of the universe (as whether or not it were self-existent as well as eternal) or about less important points. Since, besides that his expressions are oftentimes dark and ambiguous enough, and the things he delivers in several passages do not seem always very consistent, it sufficed for my purpose – which was to question vulgar notions – to examine those opinions that are by the generality of scholars taken for the Aristotelian and Peripatetic

[h] Lat. lacks this phrase.

doctrines, by which, if he be misrepresented, the blame ought to light upon his commentators and followers.

2. The rational soul or mind of man, as it is distinct from the sensitive soul, being an immaterial spirit, is a substance of so heteroclite a kind in reference to things so vastly differing from it as mere bodies are, that – since I could neither without injuring it treat of it promiscuously with the corporeal works of God, nor speak worthily of it without frequently interrupting and disordering my discourse by exceptions that would either make it appear intricate, or would be very troublesome to you or any other that you may think fit to make my reader – I thought I might, for others' ease and my own, be allowed to set aside the considerations of it in the present treatise. And the rather, because all other parts of the universe being (according to the received opinion) the works of nature, we shall not want in them subjects more than sufficiently numerous whereon to make our examen. Though I shall here consider the world but as the great system of things corporeal, as it once really was towards the close of the sixth day of the creation, when God had finished all his material works but had not yet created man.[i]

[i] Genesis 1:24–7.

SECTION II

A considering person may well be tempted to suspect that men have generally had but imperfect and confused notions concerning nature, if he but observes that they apply that name to several things, and those too such as have (some of them) very little dependence on or connection with such others. And I remember that in Aristotle's *Metaphysics*, I met with a whole chapter expressly written to enumerate the various acceptions of the Greek word φύσις, commonly rendered 'nature', of which, if I mistake not, he there reckons up six.[a] In English also we have not fewer, but rather more numerous significations of that term. For sometimes we use the word 'nature' for that author of nature whom the schoolmen harshly enough call *natura naturans* [literally, nature naturing], as when it is said that nature has made man partly corporeal and partly immaterial. Sometimes we mean by the nature of a thing the essence, or that which the schoolmen scruple not to call the 'quiddity' of a thing – namely, the attribute or attributes on whose score it is what it is, whether the thing be corporeal or not, as when we attempt to define the nature of an angel, or of a triangle, or of a fluid body as such. Sometimes we confound that which a man has by nature with what accrues to him by birth, as when we say that such a man is noble by nature, or such a child naturally forward or sickly or frightful. Sometimes we take nature for an internal principle of motion, as when we say that a stone let fall in the air is by nature carried towards the centre of the earth and, on the contrary, that fire or flame does naturally move upwards towards heaven. Sometimes we understand by nature the established course of things, as when we say that nature makes the night succeed the day; nature has made respiration necessary to the life of men. Sometimes we take nature for an aggregate of powers belonging to a body, especially a living one, as when physicians say that nature is strong or weak or spent, or that in such or such diseases nature left to herself will do the cure.[b] Sometimes we take nature for the universe or system of the corporeal works of God, as when it is said of a phoenix or a chimera that there is no such thing in nature, i.e. in the world. And sometimes too, and that most commonly, we would express by the word

---

[a]   See Book XII of Aristotle's *Metaphysics*.   [b]   Lat. lacks 'or that ... the cure'.

'nature' a semi-deity or other strange kind of being such as this discourse examines the notion of.

And besides these more absolute acceptions (if I may so call them) of the word 'nature', it has divers others (more relative), as nature is wont to be set in opposition or contradistinction to other things, as when we say of a stone when it falls downwards that it does it by a natural motion, but that if it be thrown upwards, its motion that way is violent. So chemists distinguish vitriol into natural and fictitious, or made by art, i.e. by the intervention of human power or skill. So it is said that water kept suspended in a sucking pump is not in its natural place, as that is which is stagnant in the well. We say also that wicked men are still in the state of nature, but the regenerate in a state of grace; that cures wrought by medicines are natural operations, but the miraculous ones wrought by Christ and his apostles were supernatural. Nor are these the only forms of speech that a more diligent collector than I think it necessary I should here be might instance in, to manifest the ambiguity of the word 'nature' by the many and various things it is applied to signify, though some of those already mentioned should be judged too near to be coincident. Among Latin writers, I found the acceptions of the word 'nature' to be so many, that I remember one author reckons up no less than fourteen or fifteen. From all which it is not difficult to gather, how easy it is for the generality of men, without excepting those that write of natural things, to impose upon others and themselves in the use of a word so apt to be misemployed.

On this occasion I can scarce forbear to tell you that I have often looked upon it as an unhappy thing, and prejudicial both to philosophy and physic, that the word 'nature' has been so frequently and yet so unskilfully employed, both in books and in discourse, by all sorts of men, learned and illiterate. For the very great ambiguity of this term, and the promiscuous use men are wont to make of it without sufficiently attending to its different significations, makes many of the expressions wherein they employ it (and think they do it well and truly) to be either not intelligible, or not proper, or not true – which observation, though it be not heeded, may with the help of a little attention be easily verified, especially because the term 'nature' is so often used that you shall scarce meet with any man who, if he have occasion to discourse anything long of either natural or medicinal subjects, would not find himself at a great loss if he were prohibited the use of the word 'nature' and of those phrases whereof it makes the principal part. And I confess I could heartily wish, that philosophers and other learned men (whom the rest in time would follow) would by common (though perhaps tacit) consent introduce some more significant and less ambiguous terms and expressions in the room of the too licentiously abused word 'nature' and the forms of speech that depend on it; or would at least decline the use of it as much as conveniently they can, and where they think they must employ it, would add a word or two to declare in what clear and determinate sense they use it. For without somewhat of this kind be done, men will very hardly avoid being led into divers mistakes, both of things and of one another. And such wranglings about words and names will be (if not continually multiplied) still kept on foot, as are wont to be managed with much heat, though little use, and no necessity.

And here I must take leave to complain, in my own excuse, of the scarce superable difficulty of the task that the design of a *Free Enquiry* puts me upon. For it is far more difficult than anyone that has not tried (and I do not know that any man has) would imagine, to discourse long of the corporeal works of God and especially of the operations and phenomena that are attributed to nature, and yet decline making oftentimes use of that term, or forms of speech whereof it is a main part, without much more frequent and perhaps tedious circumlocutions that I am willing to trouble you with. And therefore I hope you will easily excuse me, if – partly to shun these and to avoid using often the same words too near one another, and partly out of unwillingness to employ vulgar terms likely to occasion or countenance vulgar errors – I have several times been fain to use paraphrases or other expressions less short than those commonly received, and sometimes, for one or other of these reasons or out of inadvertence, missed of avoiding the terms used by those that admit and applaud the vulgar notion of nature: whom (I must here advertise you that), partly because they do so and partly for brevity's sake, I shall hereafter many times call naturists, which appellation I rather choose than that of naturalists because many, even of the learned among them (as logicians, orators, lawyers, arithmeticians, etc.), are not physiologers.

But if on this occasion you should be very urgent to know what course I would think expedient, if I were to propose any, for the avoiding the inconvenient use of so ambiguous a word as 'nature', I should first put you in mind that, having but very lately declared that I thought it very difficult, in physiological discourses especially, to decline the frequent [use] of that term, you are not to expect from me the satisfaction you may desire in an answer. And then I would add, that yet my unwillingness to be altogether silent, when you require me to say somewhat, makes me content to try whether the mischief complained of may not be in some measure either obviated or lessened, by looking back upon the eight various significations that were not long since delivered of the word 'nature', and by endeavouring to express them in other terms or forms of speech.

1. Instead then of the word 'nature' taken in the first sense [for *natura naturans*],^{c} we may make use of the term it is put to signify, namely God – wholly discarding an expression which, besides that it is harsh and needless and in use only among the schoolmen, seems not to me very suitable to the profound reverence we owe the divine majesty, since it seems to make the creator differ too little by far from a created (not to say an imaginary) being.

2. Instead of 'nature' in the second sense [for that on whose account a thing is what it is, and is so called], we may employ the word 'essence', which is of great affinity to it, if not of an adequate import. And sometimes also we may make use of the word 'quiddity', which, though a somewhat barbarous term, is yet frequently employed and well enough understood in the schools, and (which is more considerable) is very comprehensive and yet free enough from ambiguity.

3. What is meant by the word 'nature' taken in the third sense of it [for what belongs to a living creature at its nativity, or accrues to it by its birth] may be expressed sometimes by saying that a man or other animal is born so, and sometimes by saying that a thing has been generated such, and sometimes also that it is thus or thus qualified by its original temperament and constitution.

4. Instead of the word 'nature' taken in the fourth acception [for an internal principle of local motion], we may say sometimes that this or that body moves as it were, or else that it seems to move spontaneously (or of its own accord) upwards, downwards, etc., or that it is put into this or that motion, or determined to this or that action by the concourse of such or such (proper) causes.

^{c} The square brackets in each of the eight axioms are Boyle's.

5. For 'nature' in the fifth signification [for the established course of things corporeal], it is easy to substitute what it denotes: the established order, or the settled course of things.

6. Instead of 'nature' in the sixth sense of the word [for an aggregate of powers belonging to a body, especially a living one], we may employ the 'constitution', 'temperament', or the 'mechanism', or the 'complex of the essential properties or qualities', and sometimes the condition, the structure or the texture of that body. And if we speak of the greater portions of the world, we may make use of one or other of these terms: 'fabric of the world', 'system of the universe', 'cosmical mechanism' or the like.

7. Where men are wont to employ the word 'nature' in the seventh sense [for the universe or the system of the corporeal works of God], it is easy and as short to make use of the word 'world' or 'universe', and instead of 'the phenomena of nature', to substitute 'the phenomena of the universe' or 'of the world'.

8. And, as for the word 'nature' taken in the eighth and last of the forementioned acceptions [for either (as some pagans styled her) a goddess or a kind of semi-deity], the best way is not to employ it in that sense at all, or at least as seldom as may be, and that for divers reasons which may in due place be met with in several parts of this essay.

But though the foregoing diversity of terms and phrases may be much increased, yet I confess it makes but a part of the remedy I propose against the future mischiefs of the confused acception of the word 'nature' and the phrases grounded on it. For besides the synonymous words and more literal interpretations lately proposed, a dextrous writer may oftentimes be able to give such a form (or, as the modern Frenchmen speak, such a *tour*)[d] to his many-ways variable expressions, as to avoid the necessity of making use of the word 'nature', or sometimes so much as of those shorter terms that have been lately substituted in its place. And to all this I must add that, though one or two of the eight forementioned terms or phrases (as 'quiddity' and 'cosmical mechanism') be barbarous or ungenteel, and some other expressions be less short than the word 'nature', yet it is more the interest of philosophy to tolerate a harsh term that has been long

---

[d] Lat. lacks the parenthetical phrase.

received in the schools in a determinate sense, and bear with some paraphrastical expressions, than not to avoid an ambiguity that is liable to such great inconveniences as have been lately, or may be hereafter, represented.[e]

There are, I know, some learned men who (perhaps being startled to find nature usually spoken of so much like a kind of goddess) will have the nature of every thing to be only the law that it receives from the creator, and according to which it acts on all occasions. And this opinion seems much of kin to, if not the same with, that of the famous Helmont, who justly rejecting the Aristotelian tenent of the contrariety or hostility of the elements, will have every body, without any such respect, to act that which it is commanded to act. And indeed this opinion about nature, though neither clear nor comprehensive enough, seems capable of a fair construction. And there is oftentimes some resemblance between the orderly and regular motions of inanimate bodies and the actions of agents that, in what they do, act conformably to laws. And even I sometimes scruple not to speak of the laws of motion and rest that God has established among things corporeal, and now and then (for brevity's sake or out of custom) to call them, as men are wont to do, the laws of nature, having in due place declared in what sense I understand and employ these expressions.

But to speak strictly (as becomes philosophers in so weighty a matter), to say that the nature of this or that body is but the law of God prescribed to it, is but an improper and figurative expression. For besides that this gives us but a very defective idea of nature, since it omits the general fabric of the world and the contrivances of particular bodies, which yet are as well necessary as local motion itself to the production of particular effects and phenomena – besides this (I say) and other imperfections of this notion of nature that I shall not here insist on, I must freely observe that, to speak properly, a law being but a notional rule of acting according to the declared will of a superior, it is plain that nothing but an intellectual being can be properly capable of receiving and acting by a law. For if it does not understand, it cannot know what the will of the legislator is, nor can it have any intention to accomplish it, nor can it act with regard to it, or know when it does, in acting, either conform to it or deviate from it. And it is intelligible to me

---

[e]   The rest of this section was written *c*. 1680; see BP 14, fol. 114.

that God should at the beginning impress determinate motions upon the parts of matter, and guide them as he thought requisite for the primordial constitution of things; and that ever since, he should by his ordinary and general concourse maintain those powers which he gave the parts of matter to transmit their motion thus and thus to one another. But I cannot conceive how a body devoid of understanding and sense, truly so called, can moderate and determine its own motions, especially so as to make them conformable to laws that it has no knowledge or apprehension of. And that inanimate bodies, how strictly soever called natural, do properly act by laws, cannot be evinced by their sometimes acting regularly and, as men think, in order to determinate ends, since in artificial things we see many motions very orderly performed, and with a manifest tendency to particular and predesigned ends – as in a watch, the motions of the spring, wheels, and other parts are so contemperated and regulated that the hand upon the dial moves with a great uniformity, and seems to moderate its motion so as not to arrive at the points that denote the time of the day either a minute sooner, or a minute later, than it should do to declare the hour. And when a man shoots an arrow at a mark so as to hit it, though the arrow moves towards the mark as it would if it could and did design to strike it, yet none will say that this arrow moves by a law, but by an external though well-directed impulse.

SECTION III

But possibly the definition of a philosopher may exempt us from the perplexities to which the ambiguous expressions of common writers expose us. I therefore thought fit to consider, with a somewhat more than ordinary attention, the famous definition of nature that is left us by Aristotle, which I shall recite rather in Latin than in English – not only because it is very familiarly known among scholars in that language, but because there is somewhat in it that (I confess) seems difficult to me to be without circumlocution rendered intelligibly in English: *Natura*, says he, *est principium et causa motus et quietis ejus, in quo inest, primo per se, et non secundum accidens.*[1] [Nature is the principle and cause of movement and rest in the thing to which it belongs primarily, in virtue of itself, and not contingently.] But though, when I considered that according to Aristotle the whole world is but a system of the works of nature, I thought it might well be expected that the definition of a thing, the most important in natural philosophy, should be clearly and accurately delivered.

Yet to me, this celebrated definition seemed so dark, that I cannot brag of any assistance I received from it towards the framing of a clear and satisfactory notion of nature. For I dare not hope, that what as to me is not itself intelligible should make me understand what is to be declared or explicated by it. And when I consulted some of Aristotle's interpreters upon the sense of this definition, I found the more considerate of them so puzzled with it, that their discourses of it seemed to tend rather to free the maker of it from tautology and self-contradiction, than to manifest that the definition itself is good and instructive, and such as affords a fair account of the thing defined. And indeed, though the immoderate veneration they cherish for their master engages them to make the best they can of the definition given by him, even when they cannot justify it without strained interpretations, yet what every one seems to defend in gross, almost every one of them censures in parcels – this man attacking one part of the definition, and that man another, with objections so weighty (not to call some of them so unanswerable) that if I had no other arguments to urge against it, I might borrow enough from the commentators on it to justify my dislike of it.

[1]   2 Phys. c. I. l. 3. [Aristotle, *Physica*, II. 192b 20–2.]

However, we may hereafter have occasion to consider some of the main parts of this definition, and in the meanwhile it may suffice that we observe that several things are commonly received as belonging to the idea or notion of nature that are not manifestly, or not at all comprehended in, this Aristotelian definition, which does not declare whether the principle or cause (which expression already makes the sense doubtful) here mentioned is a substance or an accident; and if a substance, whether corporeal or immaterial. Nor is it clearly contained in this definition, that nature does all things most wisely and still acts by the most compendious ways without ever missing of her end, and that she watches against a vacuum for the welfare of the universe – to omit divers other things that you will find ascribed to her in the following section.

That the great shortness of this third section may not make it too disproportionate in length to the others this tract consists of, I shall in this place (though I doubt it be not the most proper that could be chosen) endeavour to remove betimes the prejudice that some divines and other pious men may perhaps entertain upon the account (as they think) of religion, against the care I take to decline the frequent use of that word 'nature' in the vulgar notion of it, reserving to another and fitter place some other things that may relate to the theological scruples – if any occur to me – that our *Free Enquiry* may occasion.

The philosophical reason that inclines me to forbear as much as conveniently I can the frequent use of the word 'nature' and the forms of speech that are derived from it, is that it is a term of great ambiguity – on which score I have observed that, being frequently and unwarily employed, it has occasioned much darkness and confusion in many men's writings and discourses. And I little doubt, but that others would make the like observations, if early prejudices and universal custom did not keep them from taking notice of it.

Nor do I think myself obliged, by the just veneration I owe and pay religion, to make use of a term so inconvenient to philosophy. For I do not find that for many ages the Israelites, that then were the only people and church of God, made use of the word 'nature' in the vulgar notion of it. Moses, in the whole history of the creation, where it had been so proper to bring in this first of second causes, has not a word of nature. And whereas philosophers presume that she, by her plastic power and skill, forms plants and animals out of the universal matter, the divine historian[a] ascribes the formation of them to God's immediate fiat. Genesis 1:11: 'And God said, Let the earth bring forth grass, and the herb yielding seed, and the fruit tree yielding fruit after his kind,' etc. And again, verse 24, 'God said, Let the earth bring forth the living creature after its kind,' etc., [and] verse 25, 'And God (without any mention of nature) made the beast of the earth after his kind.' And I do not remember that, in the Old Testament, I have met with any one Hebrew word that properly signifies nature in the sense we take it in. And it seems that our English translators of the Bible were not more fortunate in that than I, for having purposely consulted a late concordance, I found not that word 'nature' in any text of the Old Testament. So likewise, though Job, David, and Solomon and other Israelite writers do on divers occasions many times mention the corporeal works of God, yet they do not take notice of nature, which our philosophers would have his great vicegerent in what relates to them. To which, perhaps it may not be impertinent to add, that though the late famous Rabbi Menasseh Ben Israel has purposely written a book of numerous problems touching the creation,[b] yet I do not remember that he employs the word 'nature' in the received notion of it to give an account of any of God's mundane creatures. And when St Paul himself, who was no stranger to the heathen learning, writing to the Corinthians, who were Greeks, speaks of the production of corn out of seed sown, he does not attribute the produced body to nature; but when he had spoken of a 'grain of wheat or some other seed' put into the ground, he adds, that 'God gives it such a body as he pleaseth, and to every seed its own body', i.e. the body belonging to its kind.[2] And a greater than St Paul, speaking of the gaudiness of the lilies (or, as some will have it, 'tulips'), uses this expression, 'If God so clothe the grass of the field', etc., Matthew 6:28–30. The celebrations that David, Job and other holy Hebrews mentioned in the Old Testament make on occasion of the admirable works they contemplated in the universe are addressed directly to God himself, without taking notice of nature. Of this I could multiply instances, but shall here for brevity's sake be contented to

---

[2] 1 Corinthians 15:37, 38.

[a] i.e. the author of Genesis.

[b] *De creatione problemata XXX* (Amsterdam, 1635), by Menasseh ben Joseph, Ben Israel (1604–57), a rabbi from Amsterdam who appealed to Oliver Cromwell to readmit Jews into England, in which connection he spent two years in London from 1655. Boyle cites this book in the next section.

name a few, taken from the book of Psalms alone. In the hundredth of those hymns, the penman of it makes this: 'That God has made us' the ground of an exhortation, 'to enter into his gates with thanksgiving, and into his courts with praise', Psalms 100:3, 4.[c] And in another, 'Let the heaven and earth praise God' [that is, give men ground and occasion to praise him], congruously to what David elsewhere says to the great creator of the universe: 'All thy works shall praise thee, O Lord, and thy saints shall bless thee', Psalms 145:10. And in another of the sacred hymns, the same royal poet says to his maker, 'Thou hast covered me in my mother's womb. I will praise thee, for I am fearfully and wonderfully made, marvellous are thy works, and that my soul knoweth right well', Psalms 139:13, 14.

I have sometimes doubted whether one may not on this occasion add, that if men will need take in a being subordinate to God for the management of the world, it seems more consonant to the holy scripture to depute angels to that charge than nature. For I consider, that as to the celestial part of the universe, in comparison of which the sublunary is not perhaps the ten-thousandth part, both the heathen Aristotelians and the school philosophers among the Christians teach the celestial orbs to be moved or guided by intelligences or angels. And as to the lower or sublunary world, besides that the holy writings teach us that angels have been often employed by God for the government of kingdoms (as is evident out of the book of Daniel) and the welfare and punishment of particular persons, one of those glorious spirits is in the Apocalypse expressly styled the 'angel of the waters',[3] which title divers learned interpreters think to be given him because of this charge or office to oversee and preserve the waters. And I remember that in the same book there is mention made of an angel that had 'power', 'authority' or 'jurisdiction',[4] (ἐξουσιαν) 'over the fire'. And though the excellent Grotius gives another conjecture of the title given the 'angel of the waters', yet in his notes upon the next verse save one,[5] he teaches that there was an angel appointed to preserve the souls that were kept under the altar there mentioned. And if we take the angel of the waters

---

[3] Revelation 16:5.   [4] Revelation 14:18.

[5] Verse 7. [Citing the section on Revelation from *Annotationes in Novum Testamentum, pars tertia ac ultim* (Paris, 1650) by the Dutch theologian and jurist Hugo Grotius (1583–1645).]

[c] Here we have corrected the original text, which erroneously cites Psalm 129:34. The square brackets in the next sentence are Boyle's.

to be the guardian or conserver of them (perhaps as the Romans, in whose empire St John wrote, had special officers to look to their aqueducts and other waters), it may not be amiss to observe upon the by, that he is introduced praising his and his fellow spirits' great creator – which is an act of religion that for ought I know, none of the naturists, whether pagan or even Christians, ever mentioned their nature to have performed.

I know it may on this occasion be alleged that *subordinata non pugnant* [subordinates do not fight (against their master)], and nature being God's vicegerent, her works are indeed his. But that he has such a vicegerent, it is one of the main businesses of this discourse to call in question, and until the affirmative be solidly proved (nay, and though it were so), I hope I shall be excused, if with Moses, Job and David, I call the creatures I admire in the visible world 'the works of God' (not of nature), and praise rather him than her for the wisdom and goodness displayed in them, since among the Israelites, till they were overrun and corrupted by idolatrous nations, there was for many ages a deep silence of such a being as we now call nature. And I think it much more safe and fit to speak as did those who for so long a time were the peculiar people of God, than with the heathen poets and philosophers, who were very prone to ascribe divinity to his creatures and sometimes even to their own.

I mention these things, not with design to engage in the controversy about the authority or use of the scripture in physical speculations, but to obviate or remove a prejudice that (as I formerly intimated) I fear may be taken up, upon the account of theology or religion, against my studiously infrequent employing the word 'nature' in the vulgar sense of it; by showing that, whether or no the scriptures be not designed to teach us higher and more necessary truths than those that concern bodies and are discoverable by the mere light of reason, both its expressions and its silence give more countenance to our hypothesis than to that of the naturists.

SECTION IV

Having shown that the definition given of nature by Aristotle himself, as great a logician as he was, has not been able to satisfy so much as his interpreters and disciples what his own idea of nature was, it would be to little purpose to trouble you and myself with enquiring into the definitions and disputes of other Peripatetics about so obscure and perplexed a subject, especially since it is not my business in this tract solicitously to examine what Aristotle thought nature to be, but what is to be thought of the vulgarly received notion of nature. And though of this the schools have been the chief propagators, for which reason it was fit to take notice of their master Aristotle's definition, yet the best way I know to investigate the commonly received opinion of nature is to consider what *effata* or axioms do pass for current about her, and what titles and epithets are unanimously given her, both by philosophers and other writers and by the generality of men that have occasion to discourse of her and her actings. Of these axioms and epithets, the principal seem to be these that follow.

Natura est sapientissima, adeoque opus Naturæ est opus Intelligentiæ. [Nature is a most wise being, and thus the work of Nature is the work of Intelligence.]

Natura nihil facit frustra.[1] [Nature does nothing in vain.]

Natura fine suo nunquam excidit. [Nature never misses her own goal.]

Natura semper facit quod optimum est.[2] [Nature always does that which is best.]

Natura semper agit per vias brevissimas. [Nature always does things by the most efficient means.]

Natura neque redundat in superfluis, neque deficit in necessariis. [Nature neither uses superfluous means nor lacks necessary means.]

Omnis Natura est conservatrix sui. [All nature preserves itself.]

Natura est morborum medicatrix. [Nature is the curer of diseases.]

Natura semper invigilat conservationi universi. [Nature always looks out for the preservation of the universe.]

Natura vacuum horret. [Nature abhors a vacuum.]

From all these particulars put together, it may appear that the vulgar

---

[1] Aristotle, *De caelo*, ii. 11.

[2] Aristotle, *De caelo*, ii. 5, and *De generatione et corruptione*, ii. 10. 22. [Lat. lacks this note.]

notion of nature may be conveniently enough expressed by some such description as this. Nature is a most wise being that does nothing in vain, does not miss of her ends, does always that which (of the things she can do) is best to be done, and this she does by the most direct or compendious ways, neither employing any things superfluous, nor being wanting in things necessary; she teaches and inclines every one of her works to preserve itself. And, as in the microcosm (man) it is she that is the curer of diseases, so in the macrocosm (the world) for the conservation of the universe she abhors a vacuum, making particular bodies act contrary to their own inclinations and interests to prevent it for the public good.

What I think of the particulars that make up this panegyrical description of nature will (God permitting) be told you in due place, my present work being only to make you the clearest representation I can of what men generally (if they understand themselves) do, or with congruity to the axioms they admit and use, ought to conceive nature to be.

It is not unlike that you may expect or wish, that on this occasion I should propose some definition or description of nature as my own. But declining (at least at present) to say anything dogmatically about this matter, I know not whether I may not on this occasion confess to you that I have sometimes been so paradoxical, or (if you please) so extravagant, as to entertain as a serious doubt what I formerly intimated, viz. Whether nature be a thing or a name? I mean, whether it be a real existent being, or a notional entity somewhat of kin to those fictitious terms that men have devised that they might compendiously express several things together by one name, as when (for instance) we speak of the concocting faculty ascribed to animals. Those that consider and are careful to understand what they say, do not mean I know not what entity that is distinct from the human body, as it is an engine curiously contrived and made up of stable and fluid parts. But observing an actuating power and fitness in the teeth, tongue, spittle, fibres and membranes of the gullet and stomach, together with the natural heat, the ferment (or else the menstruum)[a] and some other agents, by their co-operation, to cook or dress the aliments and change them into chyle – observing these things, I say, they thought it convenient, for brevity's

---

[a]   The opening parenthesis is missing; we have placed it where we think it ought to be.

sake, to express the complex of those causes and the train of their actions by the summary appellation of 'concocting faculty'.

While I was indulging myself in this kind of ravings, it came into my mind that the naturists might demand of me how, without admitting their notion, I could give any tolerable account of those most useful forms of speech, which men employ when they say that 'nature does this or that', or that 'such a thing is done by nature', or 'according to nature', or else happens 'against nature'? And this question I thought the more worth answering because these phrases are so very frequently used by men of all sorts, as well learned as illiterate, that this custom has made them be thought not only very convenient, but necessary;[b] insomuch that I look upon it as none of the least things that has procured so general a reception to the vulgar notion of nature that these ready and commodious forms of speech suppose the truth of it.

It may therefore, in this place, be pertinent to add that such phrases as that 'nature' or 'faculty' or 'suction' 'does this or that' are not the only ones wherein I observe that men ascribe to a notional thing that which indeed is performed by real agents. As, when we say that the law punishes murder with death, that it protects the innocent, releases a debtor out of prison when he has satisfied his creditors (and the ministers of justice) on which or the like occasions, we may justly say that it is plain that the law – which, being in itself a dead letter, is but a notional rule – cannot in a physical sense be said to perform these things; but they are really performed by judges, officers, executioners and other men acting according to that rule. Thus when we say that 'custom does this or that', we ought to mean only that such things are done by proper agents acting with conformity to what is usual (or customary) on such occasions. And to give you a yet more apposite instance, do but consider how many events are wont to be ascribed to fortune or chance, and yet fortune is in reality no physical cause of anything (for which reason probably it is, that ancienter naturalists than Aristotle, as himself intimates, take no notice of it when they treat of natural causes), and only denotes that those effects that are ascribed to it were produced by their true and proper agents, without intending to produce them – as, when a man shoots at a deer, and the arrow lightly glancing upon the beast, wounds some man that lay beyond him, unseen

---

[b] The rest of this sentence and the next two paragraphs were written *c.* 1680; see BP 18, fols. 112–13 and 117-18.

by the archer. It is plain that the arrow is a physical agent that acts by virtue of its fabric and motion in both these effects, and yet men will say that the slight hurt it gave the deer was brought to pass according to the course of 'nature' because the archer designed to shoot the beast, but the mortal wound it gave the man happened by 'chance', because the archer intended not to shoot him or any man else. And, whereas divers of the old atomical philosophers, pretending (without good reason, as well as against piety) to give an account of the origin of things without recourse to a deity, did sometimes affirm the world to have been made by nature, and sometimes by fortune, promiscuously employing those terms, they did it (if I guess aright) because they thought neither of them to denote any true and proper physical cause, but rather certain conceptions that we men have of the manner of acting of true and proper agents. And therefore, when the Epicureans taught the world to have been made by chance, it is probable that they did not look upon chance as a true and architectonic cause of the system of the world, but believed all things to have been made by the atoms, considered as their conventions and concretions into the sun, stars, earth and other bodies, were made without any design of constituting those bodies.

While this vein of framing paradoxes yet continued, I ventured to proceed so far as to question whether one may not infer from what has been said, that the chief advantage a philosopher receives from what men call nature be not that it affords them on divers occasions a compendious way of expressing themselves. Since (thought I), to consider things otherwise than in a popular way, when a man tells me that 'nature does such a thing', he does not really help me to understand or to explicate how it is done. For it seems manifest enough that whatsoever is done in the world, at least wherein the rational soul intervenes not, is really effected by corporeal causes and agents, acting in a world so framed as ours is according to the laws of motion settled by the omniscient author of things. When a man knows the contrivance of a watch or clock by viewing the several pieces of it and seeing how, when they are duly put together, the spring or weight sets one of the wheels a-work, and by that another, until by a fit consecution of the motions of these and other parts, at length the index comes to point at the right hour of the day; the man, if he be wise, will be well enough satisfied with this knowledge of the cause of the proposed effect, without troubling himself to examine whether a notional philosopher will call the time-measuring instrument an *ens per se* [a thing in itself] or an *ens per accidens* [a variable attribute contingent on something else], and whether it performs its operations by virtue of an internal principle, such as the spring of it ought to be, or of an external one, such as one may think the appended weight. And, as he that cannot by the mechanical affections of the parts of the universal matter explicate a phenomenon will not be much helped to understand how the effect is produced by being told that nature did it, so, if he can explain it mechanically, he has no more need to think or (unless for brevity's sake) to say that nature brought it to pass than he that observes the motions of a clock has to say that it is not the engine, but it is art, that shows the hour. Whereas, without considering that general and uninstructive name, he sufficiently understands how the parts that make up the engine are determined by their construction and the series of their motions to produce the effect that is brought to pass.

When the lower end of a reed being dipped, for instance, in milk or water, he that holds it does cover the upper end with his lips and fetches his breath, and hereupon the liquor flows into his mouth. We are told that nature raises it to prevent a vacuum, and this way of raising it is called suction, but when this is said, the word 'nature' does but furnish us with a short term to express a concourse of several causes, and so does in other cases but what the word 'suction' does in this. For neither the one nor the other helps us to conceive how this seemingly spontaneous ascension of a heavy liquor is effected – which they that know that the outward air is a heavy fluid, and gravitates or presses more upon the other parts of the liquor than the air contained in the reed (which is rarefied by the dilation of the sucker's thorax) does upon the included part of the surface, will readily apprehend that the smaller pressure will be surmounted by the greater, and consequently yield to the ascension of the liquor, which is by the prevalent external pressure impelled up into the pipe, and so into the mouth (as I, among others, have elsewhere fully made out[c]). So that, according to this doctrine, without recurring to nature's care to prevent a vacuum, one that had never heard of the Peripatetic notions of nature or of suction might very well understand the mentioned phenomenon. And if afterwards he should be made acquainted with the received opinions and forms of

---

[c] Clearly a reference to various experiments Boyle had described in *New Experiments Physico-Mechanical, Touching the Spring of the Air and its Effects* (Oxford, 1660).

speech used on this occasion, he would think that so to ascribe the effect to nature is needless, if not also erroneous, and that the common theory of suction can afford him nothing but a compendious term to express at once the concourse of the agents that make the water ascend.

How far I think these extravagant reasonings may be admitted, you will be enabled to discern by what you will hereafter meet with relating to the same subjects in the seventh section of this discourse. And therefore, returning now to the rise of this digression – namely, that it is not unlike you may expect I should, after the vulgar notion of nature that I lately mentioned without acquiescing in it, substitute some definition or description of nature as mine – I hope you will be pleased to remember that the design of this paper was to examine the vulgar notion of nature, not propose a new one of my own. And indeed the ambiguity of the word is so great, and it is even by learned men usually employed to signify such different things, that without enumerating and distinguishing its various acceptions, it were very unsafe to give a definition of it, if not impossible to deliver one that would not be liable to censure. I shall not therefore presume to define a thing of which there is yet no settled and stated notion agreed on among men.

And yet, that I may as far as I dare comply with your curiosity, I shall tell you that if I were to propose a notion as less unfit than any I have met with to pass for the principal notion of nature, with regard to which many axioms and expressions relating to that word may be not inconveniently understood, I should distinguish between the universal and the particular nature of things. And of universal nature, the notion I would offer should be some such as this: that nature is the aggregate of the bodies that make up the world, framed as it is, considered as a principle by virtue whereof they act and suffer according to the laws of motion prescribed by the author of things. Which description may be thus paraphrased: that nature, in general, is the result of the universal matter or corporeal substance of the universe, considered as it is contrived into the present structure and constitution of the world, whereby all the bodies that compose it are enabled to act upon, and fitted to suffer from, one another, according to the settled laws of motion. I expect that this description will appear prolix and require to be heedfully perused, but the intricateness and importance of the subject hindered me from making it shorter, and made me choose rather to presume upon your attention than not endeavour to express myself intelligibly and warily about a subject of such moment. And this will make way for the other (subordinate) notion that is to attend the former description: since the particular nature of an individual body consists in the general nature applied to a distinct portion of the universe. Or rather, supposing it to be placed, as it is, in a world framed by God like ours, it consists in a convention of the mechanical affections (such as bigness, figure, order, situation, contexture and local motion) of its parts (whether sensible or insensible), convenient and sufficient to constitute in, or to entitle to, its particular species or denominations, the particular body they make up, as the concourse of all these is considered as the principle of motion, rest and changes in that body.

If you will have me give to these two notions more compendious expressions, now that by what has been said (I presume) you apprehend my meaning, I shall express what I called 'general nature' by 'cosmical mechanism' – that is, a comprisal of all the mechanical affections (figure, size, motion, etc.) that belong to the matter of the great system of the universe. And to denote the nature of this or that particular body, I shall style it the 'private', the 'particular', or (if you please) the 'individual mechanism' of that body, or, for brevity's sake, barely the 'mechanism' of it[d] – that is, the 'essential modification', if I may so speak, by which I mean the comprisal of all its mechanical affections convened in the particular body, considered as it is determinately placed in a world so constituted as ours is.

It is like you will think it strange that in this description I should make the present fabric of the universe a part, as it were, of the notion I frame of nature, though the generality of philosophers as well as other men speak of her as a plastic principle of all the mundane bodies, as if they were her effects; and therefore they usually call them the works of nature, and the changes that are observed in them the phenomena of nature. But for my part, I confess, I see no need to acknowledge any architectonic being, besides God, antecedent to the first formation of the world.

The Peripatetics, whose school either devised or mainly propagated the received notion of nature, conceiving (not only matter, but) the world to be eternal, might look upon it as the province, but could not as the work of nature, which in their hypothesis is its guardian without having been its architect. The Epicureans themselves, that would refer

d   Lat. lacks 'or, for brevity's ... of it'.

all things that are done in the world to nature, cannot according to their principles make what they now call nature to have been antecedent to the first formation of our present world. For according to their hypothesis, while their numberless atoms wildly roved in their infinite vacuity, they had nothing belonging to them but bigness, figure and motion; and it was by the coalition or convention of these atoms that the world had its beginning. So that according to them, it was not nature but chance that framed the world – though afterwards this original fabric of things does, by virtue of its structure and the innate and unlosable motive power of atoms, continue things in the same state for the main. And this course, though casually fallen into and continued without design, is that which, according to their hypothesis, ought to pass for nature.

And as mere reason does not oblige me to acknowledge such a nature as we call in question, antecedent to the origin of the world, so neither do I find that any revelation contained in the holy scriptures clearly teaches that there was then such a being. For in the history of the creation, it is expressly said that 'In the beginning God made the heavens and the earth',[e] and in the whole account that Moses gives of the progress of it, there is not a word of the agency of nature. And at the latter end, when God is introduced as making a review of all the parts of the universe, it is said that 'God saw everything that he had made',[3] and it is soon after added that 'He blessed and sanctified the seventh day, because in it' (or rather 'just before it', as I find the Hebrew particle elsewhere used) 'He had rested from all his works, which God created and made.'[4] And though there be a passage in the book of Job[5] that probably enough argues the angels (there called the 'sons of God') to have existed either at the beginning of the first day's work or some time before it, yet it is not there so much as intimated that they were co-operators with their maker in the framing of the world, of which they are represented as spectators and applauders, but not so much as instruments. But since revelation, as much as I always reverence it, is (I confess) a foreign principle in this philosophical enquiry, I shall waive it here and tell you that, when I consult only the light of reason, I am

---

[3] Genesis 1:31.   [4] Genesis 2:3.   [5] Job 38:4, 6, 7.

[e] Genesis 1:1.

inclined to apprehend the first formation of the world after some such manner as this.

I think it probable (for I would not dogmatise on so weighty and so difficult a subject) that the great and wise author of things did, when he first formed the universal and undistinguished matter into the world, put its parts into various motions whereby they were necessarily divided into numberless portions of differing bulks, figures and situations, in respect of each other. And that, by his infinite wisdom and power, he did so guide and overrule the motions of these parts at the beginning of things, as that (whether in a shorter or a longer time, reason cannot well determine) they were finally disposed into that beautiful and orderly frame we call the world; among whose parts some were so curiously contrived as to be fit to become the seeds or seminal principles of plants and animals. And I further conceive that he settled such laws or rules of local motion among the parts of the universal matter, that by his ordinary and preserving concourse the several parts of the universe, thus once completed, should be able to maintain the great construction, or system and economy, of the mundane bodies and propagate the species of living creatures. So that according to this hypothesis, I suppose no other efficient [cause] of the universe but God himself, whose almighty power, still accompanied with his infinite wisdom, did at first frame the corporeal world according to the divine ideas, which he had, as well most freely as most wisely, determined to conform them to. For I think it is a mistake to imagine (as we are wont to do) that what is called the nature of this or that body is wholly comprised in its own matter and its (I say not substantial, but) essential form, as if from that, or these only, all its operations must flow. For an individual body, being but a part of the world and encompassed with other parts of the same great automaton, needs the assistance or concourse of other bodies (which are external agents) to perform divers of its operations and exhibit several phenomena that belong to it. This would quickly and manifestly appear if, for instance, an animal or an herb could be removed into those imaginary spaces the schoolmen tell us of beyond the world, or into such a place as the Epicureans fancy their *intermundia* [spaces between the worlds], or empty intervals between those numerous worlds their master dreamed of.[f] For whatever the structures of

---

[f] Epicurus (341–270 BC) held that several worlds could exist simultaneously, separated by empty space.

these living engines be, they would as little without the co-operations of external agents (such as the sun, ether, air, etc.) be able to exercise their functions, as the great mills commonly used with us would be to grind corn without the assistance of wind or running water. Which may be thought the more credible, if it be considered that by the mere exclusion of the air (though not of light or the earth's magnetical effluvia, etc.) procured by the air pump, bodies placed in an extraordinary large glass will presently come into so differing a state that warm animals cannot live in it, nor flame (though of pure spirit of wine) burn, nor syringes draw up water, nor bees or such winged insects fly, nor caterpillars crawl, nay, nor fire run along a train of dried gunpowder – all which I speak upon my own experience.

According to the foregoing hypothesis, I consider the frame of the world already made as a great and, if I may so speak, pregnant automaton, that like a woman with twins in her womb, or a ship furnished with pumps, ordnance, etc. is such an engine as comprises or consists of several lesser engines. And this compounded machine, in conjunction with the laws of motion freely established and still maintained by God among its parts, I look upon as a complex principle, whence results the settled order or course of things corporeal. And that which happens according to this course may, generally speaking, be said to come to pass 'according to nature' or to be 'done by nature', and that which thwarts this order may be said to be 'preternatural' or 'contrary to nature'. And indeed though men talk of nature as they please, yet whatever is done among things inanimate, which make incomparably the greatest part of the universe, is really done but by particular bodies acting on one another by local motion, modified by the other mechanical affections of the agent, of the patient, and of those other bodies that necessarily concur to the effect or the phenomenon produced.

N.B. Those that do not relish the knowledge of the opinions and rights of the ancient Jews and heathens may pass on to the next or fifth section, and skip the whole following excursion comprised between double parentheses,[g] which, though neither impertinent nor useless to the scope of this treatise, is not absolutely necessary to it.

[In the foregoing (third) section of this treatise, I hope I have given a sufficient reason of my backwardness to make frequent use of the word 'nature'. And now in this (fourth) section, having laid down such a

[g]   i.e. square brackets.

description of nature as shows that her votaries represent her as a goddess or at least a semi-deity, it will not be improper in this place to declare some of the reasons of my dissatisfaction with the notion or thing itself, as well as with the use of the name, and to show why I am not willing to comply with those many that would impose it upon us as very friendly to religion. And these reasons I shall the rather propose, because not only the generality of other learned men (as I just now intimated) but that of divines themselves, for want of information or for some other cause, seem not to have well considered so weighty a matter.

To manifest therefore the malevolent aspect that the vulgar notion of nature has had, and therefore possibly may have, on religion, I think fit in a general way to premise what things they are which seem to me to have been the fundamental errors that misled the heathen world, as well philosophers as others. For if I mistake not, the looking upon merely corporeal and oftentimes inanimate things as if they were endowed with life, sense and understanding, and the ascribing to nature and some other beings (whether real or imaginary) things that belong but to God, have been some (if not the chief) of the grand causes of the polytheism and idolatry of the Gentiles.

The most ancient idolatry (taking the word in its laxer sense), or at least one of the earliest, seems to have been the worship of the celestial lights, especially the sun and moon: that kind of *aboda zara*, עבדדה זרה (as the Jewish writers call strange or false worships), being the most natural, as having for its objects glorious bodies, immortal, always regularly moved and very beneficial to men. There is recorded in the holy scripture a passage of Job, who is probably reputed to be at least as ancient as Moses, which seems to argue that this worship of the two great luminaries was practised in his time and looked upon as criminal by religious men and, as our English version renders the Hebrew words, 'punishable by the civil magistrate'. 'If', says Job, 'I beheld the sun when it shined or the moon walking in brightness, and my heart hath been secretly enticed, or my mouth hath kissed my hand', etc., Job 31:26, 27. And that this idolatry was practised in Moses's time may be gathered from that passage in Deuteronomy, 'And lest thou lift up thine eyes unto heaven, and when thou seest the sun and the moon and the stars, even all the host of heaven, shouldst be driven to worship them, and serve them', etc., Deuteronomy 4:19.

The Sabeans (or as many critics call them, the Zabians) are by some very learned men thought to have been the earliest idolaters. And the ablest of the Jewish rabbis, Maimonides, makes them to be so ancient that Abraham was put to dispute against them.[6] And their superstition had so overspread the East in Moses's time, that the same Maimonides judiciously observes that divers of the ceremonial laws given to the Jews were instituted in opposition to the idolatrous opinions, magical rites and other superstitions of these Zabians. Of this, he (seconded therein by our famous Selden[h]) gives several instances, to which some are added by the learned Hottinger.[7] But this only upon the by, my purpose in mentioning these Zabians being to observe to you that they looked upon the planets, and especially the sun and moon, as gods and worshipped them accordingly, taking them for intelligent beings that had a great interest in the government of the world. This may be proved out of some Eastern writers, especially Maimonides, who in one place[8] asserts the Zabians to have adored the sun and moon and the 'host of heaven' (as the scripture styles the celestial lights)[9] as true gods. And this we shall the less wonder at if we consult another place of the same learned author,[10] where he informs the readers that these idolaters (the Zabians or Chaldeans) made statues of silver and gold, those for the sun and these for the moon, which being consecrated by certain rites and ceremonies, did invite and, as it were, attract the spirits of these stars into those shrines, whence they would speak to their worshippers, acquaint them with things profitable, and even predict to them things to come. And of some such sort of speaking images, some learned critics suppose the *Teraphim* (as the original text calls them)[11] to have been, that Laban so prized as to call them 'his gods', which it is guessed Rachel stole from her father – lest, by consulting them, he might learn what way her husband and his company had taken in their flight. And the same great rabbi, having informed his readers that he saw several

---

[6] *More Nevoch.* lib. iii. cap. 30. [*More nebuchim; sive, Liber Doctor perplexorum auctore R. Mose Majemonide* (Basle, 1629), was Johannes Buxtorfius's Latin translation of the Hebrew work, *Guide for the Perplexed* (*c.* 1190), by the great Jewish physician, philosopher and rabbi Moses ben Maimon (1135–1204), usually known as Maimonides.]

[7] *Histor. Orientalis.* lib. i. cap. 8. [*Historia orientalis quae ex variis orientalium monumentis collecta* (Tiguri, 1651), by the philologist and theologian Johann Heinrich Hottinger (1620–67).]

[8] Lib. iii. cap. 36.   [9] 2 Kings 17:16 and 2 Chronicles 33:3.

[10] *Mor. Nevoch.* lib. iii. cap. 25.

[11] Genesis 31:19 and 30. [Which speak of Laban's 'household idols' as his 'gods'.]

[h] John Selden (1584–1654), English jurist and orientalist, was author of *De diis Syris syntagmata* (London, 1617).

books of the Zabian superstition, somewhere mentions one or two that treated of speaking images. And it was perhaps from these Zabians or their disciples that Zeno, the founder of the Stoical sect, taught, as Stobæus informs us,[i] that the sun, moon and the rest of the stars were endowed with understanding and prudence. And Seneca, an eminent champion of that rigid sect,[12] reprehends Epicurus and Anaxagoras (whose disciple he was in that opinion) that they held the sun to be a burning stone, or an aggregate of casual fires, and anything rather than a god.

I am sorry I could not avoid thinking the great Hippocrates to have been involved in the great error we are speaking of, when in his book *De principiis aut carnibus*, near the beginning, I met with this passage: *Videtur sane mihi id, quod* (θερμὸν) *calidum vocamus, immortale esse, et cuncta intelligere et videre, et audire et scire omnia, tum præsentia tum futura*.[j] [It seems to me that that which we call heat is immortal, and understands all, sees all, hears all and knows all, both present and future.] According to which supposition, he presently attempts to give some such account of the origin of the world's frame as he could in a very few lines, and then spends the rest of the book in giving particular accounts, how the parts of the human body come to be framed – wherein, though I commend the attempt in general, because (without acquiescing in I know not what faculties) he endeavours to give an intelligible and particular account how things come to be performed and produced, yet I cannot but look on this book as a remarkable instance of this truth: that without having recourse to the true God, a satisfactory account cannot be given of the original or primitive production of the greater and lesser world, since so great a naturalist as Hippocrates, by the help of his idolised θερμὸν, was unable to perform this task with any satisfaction to an attentive and intelligent enquirer.

And Galen himself, who was not unacquainted with Moses's writings and lived where Christianity was propagated through a great part of the world – Galen, I say, even in that admirable treatise *De usu partium*, where he so excellently declares and celebrates the most wise author of

---

[12] Sen. de Benef. lib. vii. cap. 21. [Citing *De beneficiis* by the Roman Stoic philosopher Seneca (4 BC–AD 65).]

[i] Zeno (335–263 BC) of Citium, founder of the Stoic school of philosophy. The Greek anthologist Stobaeus (fifth century AD) wrote *Eclogarum physicarum*.

[j] *De carnibus*, attributed to Hippocrates.

things, was so far transported with the error which infected so many other heathen philosophers, that he fancied the earth itself (though he speaks contemptibly of it) had a certain soul or mind imparted to it by the superior bodies, which (he said) is so conspicuous, first in the sun, next in the moon and afterwards in the other stars, that by their beauty the contemplator will be induced to think it reasonable, that the more pure their corporeal substance is, it is inhabited by a mind so much the better and more perfect than that of these terrestrial bodies. And having spoken of the 'reasoning nature' that shined in Plato, Aristotle, Hipparchus, Archimedes, etc., he thus infers, *Si igitur in tanta colluvie (quo enim alio nomine quis appellet id quod ex carne, sanguine, pituita, ac bile utraque est conslatum) mens gignatur, adeo eximia et excellens; quantam ejusdem putandum est esse excellentiam in sole, luna, aliisque etiam sideribus?* [If so fine and excellent a mind can arise, then, in such a rag-bag as this (what other name is one to give this conjunction of flesh, blood, phlegm and two sorts of bile?), how excellent will be that which reposes in the sun, the moon and the other stars?] To which he subjoins, *Mihi quidem, dum hæc mecum voluto, non exigua quædam mens talis, per ipsum etiam nos aerem ambientem, esse extensa videtur. Fieri enim non potest, quum lucis ipsius solis sit particeps, quin vim etiam ab ipso assumat.*[13] [When I think on this matter, it seems to me that such a mind, permeating the very air that surrounds us, is far from small. It could not be so, when it shares in the sun's light: nay, draws its strength therefrom.] But this upon the by.

Nor did this opinion of the divinity of the celestial bodies die with the Zabians or the Greek philosophers. For I found by some questions I proposed to an inquisitive person,[k] who, having lived many years in China and several of the neighbouring kingdoms, had acquired skill enough in the tongues to converse with the natives – I found (I say) that in a solemn conference he had with some of the more eminent and philosophical doctors of the Chinese religion, they frankly professed that they believe the heavenly bodies to be truly divine and to be worshipped, and that upon this particular ground, that they imparted to men such good things as light, heat, rain, etc., and the productions and

[13] Galenus de usu Partium, l. xvii. apud Lacunam in Epitome Oper. Galenii. [Galen, *De usu partium corporis humani, libri xvii*, as quoted from one of the many editions prepared by Andres de Laguna (1499–1559), entitled *Epitome omnium Galeni per A. Lacunam collecta.*]

[k] Not identified.

consequences of these. And this belief, they declared, they thought more rational than that of the Europeans who worship a deity whose neither shape, nor colour, nor motion, nor efficacy on sublunary things were at all visible. It agrees very well with the opinion of the ancient Greeks who, as Origen relates,[14] called the sun, moon and the stars ἐμφανεῖς θεοὺς καὶ αἰσθητούς, 'conspicuous and sensible gods'. And we are taught by Eusebius that the ancient Egyptian theologisers, whose religion was near of kin to that of the Chaldeans, if not borrowed of it, looked upon the sun and moon, whom they worshipped under the names of Osiris and Isis, not only as the chief gods, but as the makers and governors of much (if not of all) of the rest of the universe.[15]

I will not here enquire whether these old heathen philosophers did, besides the stars and other beings that they adored as gods, believe one only *Numen* or supreme deity. But that may suffice for my present purpose, which seems manifest: viz., that they ascribed to sensible beings, attributes peculiar to the true God; that this was occasioned by their thinking them intelligent and governing; and that these inferior beings were by far the most usual and familiar objects both of their discourses and their worship; and that they did (to use the phrase of the apostle of the Gentiles) worship the 'creature besides', or 'more than' (for the Greek word παρὰ may signify either) the 'creator',[16] who by Moses, the prophets and the apostles, expressly declares a dislike of this worship, and even in that more specious and seemingly excusable kind of it which was in use among the ten tribes that professed, and perhaps believed, their worship to be directed to the one supreme God, and him the true God of Israel. But this also upon the by.

This belief that the world and divers of its principal parts, as the sun, moon, stars, etc., were animated and endowed with intelligent minds, was so contagious that not only it helped to seduce the emperor Julian

---

[14] Origen. cont. Celsum. l[iber]. v. [*Contra Celsum*, written in 248 by the theologian Origen (185–253).]

[15] Præparat. l. iii. c. 4. [*Praeparatio evangelicae*, by Eusebius (265–340), bishop of Caesarea.] *Damascius vita Isidori apud Photium: 'Colunt præ cæteris Diis Ægyptii Osirim et Isin (i.e. Solem et Lunam,) illum omnia condere, et figuris numerisque materiam adornare arbitrati.'* ['The Egyptians worship before all other gods Osiris and Isis (i.e. the sun and the moon), considering the former to be the founder of everything and to adorn matter with form and number'; from Damascius, *Vita Isidori*, quoted by St Photius I (*c.* 820–91), patriarch of Constantinople, in his *Bibliotheca*, ed. I. Bekker (Berlin, 1824), p. 335 a 30. The words in parentheses do not appear in the Greek original.]

[16] Romans 1:25.

the emperor Julian from Christianity to heathenism (in so much that he gives the sun solemn thanks for his advancement to the Roman monarchy[17]), but it infected very learned men among the Jews and Christians. Of the former, I shall need to name but two, the first being the famousest and judiciousest of the ancienter rabbis, Maimonides, in whom (I confess) I wondered to find this assertion that the sun and stars were animated beings, endowed with understanding and will.[18] And the other, being reputed the chief and the most learned of the moderns, Menasseh Ben Israel (with whom I have conversed at Amsterdam), who in his problems *De creatione*, has this notable passage: *Quod de intelligentiis tradunt id vero mera fabula est; nam cœli, secundum Rabbi Mosem, et rei veritatem, habent animas proprias rationali vita prœditas, sicut alibi a me demonstrabitur.*[19] [What they relate about intelligent beings is a mere fable. For the heavens, according to Rabbi Moses, and the truth of the matter, have their own souls, endowed with rational life. I shall demonstrate this elsewhere.] And a greater man than Maimonides, Origen himself, among the Christians, not only in one place adventures to say, *Siquidem etiam cœlestes stellæ animalia sunt rationalia, virtute prœdita, illustrata cognitionis lumine, a sapientia illa quæ est splendor æterni luminis,*[20] [Since the heavenly stars are rational animals, endowed with virtue and enlightened with the light of knowledge by that wisdom which is the Splendour of the Everlasting Light] but in another proceeds so far that I found (not without surprise) that he says, 'The Christians sing hymns to God, the Lord of all, and God the Word; no otherwise than do the sun, moon and stars, and the whole heavenly host, since all these, being a heavenly choir, do with just men celebrate the supreme God, and his only begotten [Son].'[21] The boldness of these unjustified paradoxes I the less wonder at, when I consider what has for many ages been taught by the school philosophers from Aristotle:

---

[17] '*Sed nec illam, quam eiusdem Numinis (Solis) beneficio adeptus sum, sortem conditionemque parvi facio; quod ex eo genere, penes quod Terrarum Dominatus atque Imperium est, temporibus nostris ortum acceperim.*' Julian, *Ad regem solem*. [From a hymn to the sun god by the Roman Emperor Julian the Apostate (332–63), who converted to paganism under the influence of the Neoplatonic teacher Maximus: 'I do not disdain the lot of which I was held worthy by this Godhead [the sun], namely to have been born to the line that in my own time has power and sway over the world.']

[18] *More Nevochim.* l. 3. cap. 29. (*ni fallor* [unless I am mistaken]).

[19] Pag. M[ihi]. 98. ['Page 98 in my copy.' Citing *De creatione problemata* by Menasseh ben Israel.]

[20] Origen, *Contra Celsum*, v 10.

[21] Origen, *Contra Celsum*, viii 67. [The addition '[Son]' is Boyle's.]

namely, that the celestial spheres had their peculiar intelligences, that is, rational, immortal, powerful and active beings.

It is true that in the Jews and Christians I have been speaking of, the malignity of the error they embraced was corrected and mastered by the sound and orthodox principles they held together with it. But still, it is dangerous for those that would be loyal to him that styles himself a 'jealous God'[22] to adopt premises that have been able to mislead such great persons, and from which many famous philosophers have plausibly enough drawn consequences very repugnant to true religion. Nor are Christians themselves so much out of danger of being seduced by these heathenish notions about an intelligent world, but that (not again to mention the apostate emperor) even in these times there is lately sprung up a sect of men, as well professing Christianity as pretending to philosophy, who (if I be not misinformed of their doctrine) do very much symbolise with the ancient heathens, and talk much indeed of God, but mean such a one as is not really distinct from the animated and intelligent universe, but is on that account very differing from the true God that we Christians believe and worship. And though I find the leaders of this sect to be looked upon by some more witty than knowing men as the discoverers of unheard-of mysteries in physics and natural theology, yet their hypothesis does not at all appear to me to be new, especially when I remember, besides the passages of the ancients cited in this paper, some others of the same import, such as is particularly that of Lucan.[1]

> *Estque Dei sedes, ubi terra, et pontus, et aer,*
> *Et cœlum, et virtus: superos quid quærimus ultra?*
> *Jupiter est quodcunque vides, quocunque moveris.*

> [And is the house of god where the earth, the sea, the air, the sky and virtue are; why seek we gods beyond those? Jupiter is whatever you see, wherever you go.]

The great affinity between the soul of the world, so much talked of among the heathen philosophers, and the thing that men call nature, makes it fit for me to take notice in this place of the influence which the

---

[22] Exodus 20 [verse 5].

[1] From the epic poem *De bello civili* by the Roman poet Marcus Annaeus Lucanus (39–65), nephew of Seneca. In Lat., the final line reads: '... *quodcunque vides, Jovis omnia plena*' [... whatever you see, all full of Jove].

belief of that imaginary soul had upon the Gentiles with reference to religion. That divers of the ancient philosophers held the world to be animated has been observed by more than one learned man. But that which makes more for my present purpose is, that the same old sages did also (at least for the most part) believe that this mundane soul was not barely a living, but a most intelligent and wisely active being. This may be easily enough discerned by him that shall heedfully peruse Diogenes Laertius's *Lives of the Philosophers*, and particularly of Zeno.[m] But at present I shall rather make use of an author who, though he be very seldom cited for philosophical history, seems to me to have been very well versed in it. The writer I mean is the acute sceptic Sextus Empiricus (who is thought to have lived about Plutarch's time, and by some, to have been his nephew[n]), who recites a long ratiocination of Xenophon – which, whether it be solid or not, is at least ingenious and plausible, but too prolix to be transcribed in this place, where it may suffice to say that he thus concludes:[23] *Est ergo mundus mente præditus et intelligens, etc.* [So the world has a mind and is intelligent, etc.], which assertion Sextus himself thus proposes for him: *Si non esset aliqua mens in mundo, neque ulla mens in te esset. Est autem in te mens aliqua; ergo est etiam in mundo. Et ideo mundus est mente et intelligentia præditus.* [If there were no mind in the world, there would be no mind in you. But there is a mind in you: so there is also one in the world. Thus the world has a mind and intelligence.] The same sceptic introduces Zeno Cittiens, discoursing thus: *quod immittit semen ejus quodest particeps rationis, est ipsum quoque rationis particeps. Mundus autem emittit semen ejus quod est particeps rationis; est ergo mundus rationis particeps.* [What emits the seed of a rational thing is itself rational. Now the world emits the seed of a rational thing: so the world is rational.] To which testimonies I might add many others out of the same author, who in the same discourse tells us that the Stoics held the world to be an animal.

But the opinion that the old philosophers we have been speaking of held, of the world's being endowed with an understanding or rational

---

[23] Sextus Empiricus, *Adversus mathematicos*, lib. 8. [*Adversus mathematicos* (Book IX by modern reckoning, also referred to as Book I of *Adversus physicos*), by Sextus Empiricus. Boyle cites, respectively, IX 95, 98, 101, 95 and 98.]

[m] Book VII of *De vitis philosophorum*, by Diogenes Laertius (200–50), includes Zeno.

[n] Boyle confuses the Pyrrhonist sceptic philosopher Sextus Empiricus (fl. *c.* 200 AD) with Plutarch's nephew Sextus of Chaeronea, Platonist philosopher and teacher of the emperors Marcus Aurelius and Lucius Verus.

soul, will be yet more evident by what I now proceed to allege, to manifest how this opinion of theirs led them to the worship of another than the true God. Sextus Empiricus, in the lately cited discourse of Xenophon, infers from the world's being an intelligent being that it is also a divine one. For to the lately recited conclusion, *est ergo mundus mente præditus et intelligens* [so the world has a mind and is intelligent], he immediately subjoins this other, *et ideo deus* [and is therefore a god].[24] And a little after, repeating their discourse that defended this argumentation of Xenophon against an objection, he concludes their reasoning thus: *Ideo mundus est mente et intelligentia præditus: cum sit autem mente et intelligentia præditus, est etiam deus.* [So the world has a mind and intelligence: now since it has a mind and intelligence, it is also a god.] *Quemadmodum,* says also Phurnutus the philosopher,[o] *nos anima gubernamur, sic et mundus animam habet, quæ vindicet illum ab interitu; et hæc vocatur Jupiter.* [Just as we are governed by a mind, so likewise the world has a mind which preserves it from ruin; and its name is Jupiter.] To which agrees that in Cicero's *Academic Questions*: *Mundum esse sapientem, et habere mentem, quæ se ipsam fabricata sit, et omnia moderetur, regat.*[p] [That the world is wise and has a mind that made itself, which governs and rules all things.] And the reasoning of the Stoics in St Augustine is very clear to the same purpose: *Dicunt,* says he, speaking of the embracers of that sect, *omnia sidera partes Jovis esse, et omnia vivere atque rationales animas habere, et ideo sine controversia deos esse.*[25] [They say that all the stars are parts of Jupiter, and all live and have rational souls, and thus, without a shadow of a doubt, are gods.] And Socrates is introduced by Aristophanes as no less than invocating the air and the ether together, in these words:

O Rex, O Imperator, aer vaste, quæ terram contines suspensam,
Nec non splendide æther.[q]

---

[24] P.M. 326. ['Page 326 in my copy'].

[25] Augustine, *De civitate Dei*, iv. 11. [The first edition cites Bk. VII, ch. 2.]

[o] 'Phurnutus' is another name for the Stoic philosopher Lucius Annaeus Cornutus (fl. *c.* 60 AD), author of the *Theory Concerning the Nature of the Gods*, also known as *Concerning Allegories.*

[p] Cicero, *Academics*, ii.37. The original passage reads, *mundum esse sapientem, habere mentem, quae et se et ipsum fabricata sit, et omnia moderetur, moveat, regat.* [That the world is wise, and has a mind which made both itself and the world, which governs, moves and rules all things.]

[q] From a Latin version of *Clouds* by the Athenian comic dramatist Aristophanes (*c.* 450–385 BC). The verses (lines 264–5) are the start of a prayer put in the mouth of Socrates, parodying his religious views, as (mis)understood by the Athenian in the street.

[O King, O Emperor, boundless air, who holdest the earth aloft; and you, resplendent ether!]

Which brings into my mind that plain confession of the poet Manilius:

Qua pateat, mundum divino Numine verti,
Atque ipsum esse Deum.[r]

[To me no argument seems so compellingly clear that the world is moved by a divine spirit, and is itself a god.]

To all these I shall add that notable and express passage of the elder Pliny: *Mundum et hoc quod alio nomine cœlum appellare libuit, cujus circumflexu teguntur omnia, numen esse credi parest, æternum, immensum, neque genitum, neque interiturum unquam. Sacer est, æternus, immensus, totus in toto, vero ipse totum, finitus et infinito similis, extra, intra, cuncta complexus in se, idemque Naturæ opus, et rerum ipsa Natura.*[26] [The world, and that which goes by another name, the sky, by whose arched vault all things are covered, may be regarded as a deity: eternal, immense, unborn, destined never to die. It is holy, everlasting, vast, all-encompassing, truly itself a total being, finite but like something infinite, embracing in itself everything that exists inside and outside itself, at once a work of nature and itself the very nature of things.]

If it be objected that the passages I have cited out of heathen philosophers concern the soul of the world and not nature, I answer that the affinity of these two is so great that divers of the old sages seem to have confounded them, and not to have made account of any other universal nature than the soul of the world. And however, the great and pernicious errors they were led into by the belief that the universe itself and many of its nobler parts besides men were endowed not only with life, but understanding and providence, may suffice to make us Christians very jealous of admitting such a being as that which men venerate under the name of 'nature', since they ascribe to it as many wonderful powers and prerogatives as the idolaters did to their adored mundane soul. But I shall give a further answer to the above proposed objection, if I can show how sacrilegiously they abused the being we are speaking of, as well under the very name of 'nature', as under that of the 'soul of the world'. On this occasion I remember a passage in Seneca

---

[26] Natur. Hist. l. 2. c. 1. [Pliny the Elder (23–79), *Historia naturalia.*]

[r] From the *Astronomica*, i.484–5, by the Roman astrologer and poet Manilius (fl. 0–30).

that I did not expect to meet with where, speaking of some ethnic opinions about thunder, *Non Jovem*, says he, *qualem in capitolio colimus, fulmina mittere, sed custodem rectoremque universi, animam ac spiritum mundani hujus operis dominum et artificem, cui nomen omne convenit.*[27] [They do not think that Jupiter, whom we worship on the capital, sends the thunderbolts, but that he is the guardian and director of the universe, the soul and spirit, the lord and maker of this worldly creation; for him every name is fitting.] To which, within a few lines after, he adds, *Vis illam naturam vocare? Non peccabis, est enim ex quo nata sunt omnia, cujus spiritu vivimus. Vis illam vocare mundam? Non falleris, ipse enim est totum quid, totus suis partibus inditus et se sustinens vi sua.* [Do you wish to call it nature? You will not be wrong to do so; for it is that from which all things are born, that by whose spirit we live. Do you wish to call it the world? This is no mistake: for it is itself a complete entity, wholly located in its parts and sustaining itself by its own strength.] And the same author elsewhere: *Nihil*, says he, *Natura sine Deo est, nec Deus sine Natura, sed idem est uterque.*[28] [There is no nature without God or God without nature: the two are identical.] And in another of the Roman sages we have this passage: *Natura est igitur quæ continet mundum omnem, eumque tuetur, et quidem non sine sensu ac ratione.*[s] [So nature is what contains the whole world, and looks after it, so doing not without sense and reason.] And the opinion, not of a private philosopher, but of the sect of Stoics, is thus delivered by Lactantius: *Isti uno Naturæ nomine res diversissimas comprehenderunt, Deum et mundum, artificem et opus, dicuntque alterum sine altero nihil posse, tanquam Natura sit Deus mundo permistus. Nam interdum sic confundunt, ut sit Deus ipsa mens mundi, et mundus sit corpus Dei; quasi vero simul esse cæperint mundus et Deus.*[29] [They understand a great variety of things – God and the world, the creator and his creation – by the one word 'nature', and they say that the one can do nothing without the other, as if nature were God infused throughout the world. And sometimes they confound the two, so that God is just the world's mind, and the world just God's body, as if God and the world had the same beginning.]

---

[27] *Naturales quæstiones*, ii. 45.

[28] [Seneca,] *De beneficiis*, iv. 8. 2. [The first edition cites Bk. IV, ch. 7.]

[29] Lib. 7. cap. 1. [From *Divinae Institutiones*, VII. iii.3 by the patristic writer Lactantius (*c.* 250–317).]

[s] Quoting an unidentified author.

And, to let you see that in this our *Free Enquiry*, I do not without cause here and there style nature sometimes a semi-deity and sometimes a goddess, and talk of some men's idolising her, I shall here annex part of a hymn of Orpheus's addressed immediately to nature,[t] Ὦ φυοὶ παμμήτειρα θεα, which his interpreter thus renders into Latin:

> *O Natura omnium Mater Dea, artificiosa admodum Dea,*
> *Suscitatrix honorabilis, multa creans, Divina Regina,*
> *Omnidomans, indomita gubernatrix, ubique splendens.*

[O Nature, Goddess Mother of all, most craftful Goddess, honoured Inspirer, Maker of much that is, divine Queen, All-dominating, unbowed Governess, who everywhere shinest.]

And after a few lines:

> *Ætheria, terrestris, et Marina Regina, etc.*

[Queen of Air, Earth and Sea]

I know Aristotle and his commentators do not so directly idolise nature as did Orpheus (or whoever was the ancient author of the hymns that bear his name), but yet I doubt they pass further than they can justify when they so freely and often assert that *Natura est sapientissima* [Nature is the wisest], that *Opus Naturæ fine suo nunquam excidit* [The work of nature never misses its own goal], that *Natura semper quod optimum est facit* [Nature always does that which is best] (to which may be added other like axioms).[u] And when they most commonly call the works of God the works of nature, and mention him and her together, not as a creator and a creature, but as two co-ordinate governors like the two Roman consuls, as when they say frequently and without scruple (what I find to have been first by Aristotle himself[30]) that *Deus et Natura nihil prorsus faciunt frustra* [God and nature do nothing wholly in vain] – to which phrase may agree that expression of Ovid, where, speaking of the chaos, while the bodies that composed it lay shuffled together and were not yet packed,

---

[30] Aristotle, *De caelo*, lib. ii. cap. 5. [Lat. lacks the parenthetical phrase; *'prorsus'* is missing in the English edition.]

[t] Orpheus, by some accounts a son of Apollo, was the most famous poet and musician in Greek mythology. The Orphics, a religious sect that appeared *c.* fifth century BC, attributed many hymns to Orpheus. The verses here are from hymn no. 10 in *Orpheus hymni*, ed. W. Quandt (Berlin, 1955).

[u] For the rest of the axioms Boyle has in mind, see the opening paragraph of this section.

he says, *Hanc Deus et melior litem Natura diremit.*[v] [This turmoil God, a higher form of nature, resolved.]

To the recital of the irreligious errors of the ancient heathens about the divinity of the world and some of its principal parts (as the sun, moon, stars, ether, etc.), I should add a redargution of them, if I thought it necessary in this place solemnly to refute opinions, some of which are altogether precarious, and others very improbable. Those Greek and Latin philosophers that held the sun to be a fire were much at a loss to find out fuel to maintain the flame. But those Zabians and Chaldeans that thought him endowed not only with a living soul, but with understanding and will, must (if they had duly considered things) have been much more puzzled to find not only food for so vast a body (above 160 times bigger than the terraqueous globe), but to find in him the organs necessary to the preparation and digestion of that food and to the other functions that belong to animal nutrition. And if we admit the Cartesian hypothesis, the way whereby the sun, fixed stars and planets are generated will sufficiently manifest them to be neither intelligent nor living bodies.[w] And perhaps I could here propose a quite other hypothesis about the nature of the sun and the fuel of its fire,[x] that may be countenanced by some phenomena and experiments without making him other than an igneous and altogether inanimate body, whose flame needs to be repaired by fuel furnished to it nearer [at] hand than from the sea or earth. But I purposely omit such objections against the opinion I oppose, as – though drawn from the dictates of sound philosophy about the origin of things – may be questioned without being to be cleared in few words.

It is also without proof that it is presumed and asserted that the celestial bodies newly mentioned are endowed with understanding and prudence, especially so as to be able to know the particular conditions and transactions of men, and hear and grant the prayers of their worshippers. And the moon, which was one of their principal deities, and by them preferred before all the other planets and stars (the sun excepted), is so rude and mountainous a body that it is a wonder that speculative men, who considered how many, how various and how

---

[v]   From Ovid, *Metamorphoses*, i.21.

[w] Descartes gave a mechanistic account of the origin of stars and planets in *The Principles of Philosophy* (Amsterdam, 1644), Part III, articles 54 and 146.

[x]   A 'conjectural discourse' on this topic by Boyle is listed in an eighteenth-century inventory of his manuscripts, but is now lost; see BP 36, fol. 166.

noble functions belong to a sensitive soul, could think a lump or mass of matter, so very remote from being fitly organised, should be animated and governed by a true living and sensitive soul. I know that both these deifiers of the celestial globes and also the heathen disciples of Aristotle, besides divers of the same mind, even among the Christians, say great and lofty things of the quintessential nature of the heavenly bodies and their consequent incorruptibility, of the regularity of their motions and of their divine quality of light that makes them refulgent. But the persuasion they had of this quintessential nature of the superior part of the world was not, if I guess aright, grounded upon any solid physical reason, but was entertained by them for its congruity to the opinion they had of the divinity of the celestial bodies – of which Aristotle himself, especially in his book *De Cœlo*,[31] speaks in such a way as has not a little contributed among his followers to such an excessive veneration for those bodies, as is neither agreeable to true philosophy, nor friendly to true religion. He himself takes notice that the Pythagoreans held our earth to be one of the planets, and that it moved about the sun, which they placed in the middle of the world.[32] And since this hypothesis of the earth's motion was in the last age revived by Copernicus, not only those great men Kepler, Galileo and Gassendi, but most of the best modern astronomers and, besides Descartes and his sect, many other naturalists have embraced this hypothesis – which, indeed, is far more agreeable to the phenomena, not only than the doctrine of Aristotle (who was plainly mistaken about the order and consistence of the heavens), but than the ancient and generally received Ptolemaic system. Now, supposing the terraqueous globe to be a planet, he that considers that it is but a round mass of very heterogeneous substances (as appears by the differing natures of its great constituent parts, land and sea) whose surface is very rude and uneven and its body opacous, unless as it happens to be enlightened by the sun, moon and stars, and so very inorganical for so much as nutrition that it seems wholly unfit to be a living animal, much less a rational one – I say, he that considers such things will scarce be forward to ascribe understanding and providence, much less a divine nature, to the other stars.

---

[31] See Lib. ii. cap. 3. [The first edition erroneously cites the non-existent Bk. xi.]

[32] Aristotle, *De Cœlo*, book ii, chap. 13. [The first edition erroneously cites the same non-existent book (xi). Furthermore, according to Aristotle, the Pythagoreans placed both the sun and the earth, along with all other planets, in orbit about a central fire, the 'watch-tower of Zeus'.]

As for instance, to the moon, which our best telescopes manifest to be a very craggy and mountainous body, consisting of parts of very differing textures (as appears by her brighter parts and permanent spots), and which of herself is opacous, having no manifest light but what she borrows from the sun, and perhaps from the earth.[y]

As for the boasted immutability of the heavenly bodies, besides that it may be very probably called in question by the phenomena of some (for I do not say every one) of the comets that by their parallax were found to be above the moon, and consequently in the celestial region of the world;[z] besides this, I say, the incorruptibleness and immutability of the heavenly bodies is more than probably disprovable by the sudden and irregular generation, changes and destruction of the spots of the sun – which are sometimes so suddenly destroyed that (I remember) in the year 1660 on the 8th of May, having left in the morning a spot whose motions we had long observed through an excellent telescope, with an expectation that it would last many days visible to us, we were surprised to find that when we came to observe it again in the evening, it was quite dissipated, though it seemed thick. And by comparing it to the sun, we estimated the extent of its surface to be equal to that of all Europe.

As to the constancy of the motions of the stars, if the earth (which we know to be inanimated) be a planet, it moves as constantly and regularly about the sun (in that which they call the great orb) as the other planets do, or as the moon does about the earth. And I consider that, though we should suppose our globe not to be a planet, yet there would manifestly be a constant motion, and regular enough, of a great part of it. Since (bating some anomalies that shores, winds and some other extrinsic things occasion) there is a regular ebbing and flowing twice a day, and also springtides twice a month of that vast aggregate of waters (the ocean), which perhaps is not inferior in bulk to the whole body of the moon, and whereof also vast tracts are sometimes observed to shine.

And lastly, whereas a great proof of the divinity of the stars is taken from their light, though I grant it to be the noblest of sensible qualities, yet I cannot think it a good proof of the divine or very excellent nature of bodies endowed with it, whether they be celestial or not. For whereas

---

[y] An allusion to the phenomenon of earth-shine, first seen by Galileo.

[z] An allusion to the discovery in the last quarter of the sixteenth century by Tycho Brahe (1546–1601) that the diurnal parallax of comets places their orbits above the orbit of the moon.

the Zabians and Chaldeans considered and adored the planets as the chief gods, our telescopes discover to us that, except the sun (if he be one, rather than a fixed star), they shine but by a borrowed light, in so much that Venus, as vividly luminous as it appears to the naked eye, is sometimes seen (as I have beheld it) horned like the moon in no long time after her change.[a] And at this rate also the earth, whether it be a planet or no, is a luminous body, being enlightened by the sun; and possibly, as a body forty times bigger, communicates more light to the moon than it receives from her, as is probably argued from the light seen on the surface of the moon in some of her eclipses.[b] And though in the night, when the darkness has widened the pupils of our eyes and the moon shines with an unrivalled lustre, she seems exceeding bright, yet she may be for ought I know more opacous than the solid part of the terrestrial globe. For I remember that I have more than once heedfully observed a small cloud in the west, where the moon then was, about sunset. And comparing them together, the little cloud – as opacous and loose a body as it was – reflected the light as strongly to my eye as did the moon, that seemed perhaps to be not far from it, both of them appearing like little whitish clouds. Though afterwards, as the sun descended lower and lower beneath the horizon, the moon grew more and more luminous.

And speaking of light indefinitely, it is so far from arguing a divine nature in the bodies that are endowed with it – whether as the planets, by participation from an external illuminant, or as the sun, from an internal principle – that a burned stone, witness that of Bologna,[c] will afford in proportion to its bulk incomparably more borrowed light than one of the planets. And a light from its internal constitution may be found not only in such abject creatures as insects, whether winged as the cucupias[d] of Hispaniola or creeping as our glow-worms, but also in bodies inanimate and corrupted, as in rotten wood, in stinking whitings, and divers other putrified fishes. I cannot now stay to enquire how the Zabians, and such idolators as they, could make out the connection, symmetry and subordination or dependence of the several parts of the world, composed of so many different and distant beings, endowed not

---

[a] An allusion to the phases of Venus, first seen by Galileo.
[b] Another allusion to earth-shine.
[c] 'Bologna stone', a naturally occurring form of barium sulphide, was a dense, white stone discovered near Bologna in 1603 by the Italian alchemist Vincenzo Cascariolo, who made from it a phosphorescent material originally thought to be the philosopher's stone.
[d] Probably a reference to the *cocuyo* (click beetle); some tropical varieties are luminescent.

only with animal souls, but with their distinct and peculiar under-standings and wills and many of them also with divine nature. Nor shall I consider how strange a monster, rather than an animal and a deity, those many heathen philosophers and their adherents must make of the universe, who held it to be but one, and yet were of the paradoxical opinion that (as has been elsewhere noted) is roundly professed by Stobæus at the very beginning of his *Physical Eclogues*, Ζεύς οὖν, etc., i.e., *Jupiter (quidem) totus mundus est: animal ex animalibus; Numen ex Numinibus compositum.*[e] [Jupiter is indeed the whole world: an animal formed from animals, a deity from deities.]

These (I say) and the like objections against the pagan doctrine, I must not now insist on, because I perceive that I have slipped into a somewhat long digression (which yet perhaps may not be altogether unseasonable or useless), which therefore I shall here break off to resume and conclude the discourse that this section was allotted to, which I might easily have enlarged. But I presume there is enough said in it already to let you see that it is a dangerous thing to believe other creatures than angels and men to be intelligent and rational, especially to ascribe to any of them an architectonic, provident and governing power. And though I readily acknowledge that there is no great danger that well-instructed Christians should, like some heathens, worship nature as a goddess, yet the things I formerly alleged, to show it unsafe to cherish opinions of kin to those that misled a multitude even of philosophers, make me fear too many – and not a few of the learned themselves – may have a veneration for what they call nature much greater than belongs to a mere creature: if they do not, to use a scripture expression, 'worship the creature, above' (or 'besides') the 'creator',[33] who – and not the world, nor the soul of it – is the true God. And though I should grant that the received notion of nature does neither subvert nor much endanger any principle of religion, yet that is not enough for the purpose of those naturists I reason with, since they are here supposed to make it a fault in others not to ascribe to the nature they venerate, as much as themselves do. And they represent their own notion of it not only as innocent, but as very useful, if not necessary, to religion.][f]

---

[33] Romans 1:25.

[e] See above, p. 43.

[f] Readers are reminded that this bracket marks the end of a long digression that began on p. 40.

SECTION V

I come now, Eleutherius, to acquaint you with some of the reasons that have made me backward to entertain such a notion of nature as I have hitherto discoursed of. And I shall at present comprise them under the following five.[a]

1. The first whereof is, that such a nature as we are speaking of seems to me to be either asserted or assumed without sufficient proof. And this single reason, if it be well made out, may (I think) suffice for my turn. For in matters of philosophy, where we ought not to take up anything upon trust or believe it without proof, it is enough to keep us from believing a thing, that we have no positive argument to induce us to assent to it, though we have no particular arguments against it. And if this rule be to take place in lesser cases, sure it ought to hold in this, where we are to entertain the belief of so catholic an agent that all the others are looked upon but as its instruments, that act in subordination to it; and which, being said to have an immediate agency in many of the phenomena of the world, cannot but be supposed to be demonstrable by divers of them. I have yet met with no physical arguments, either demonstrative or so much as considerably probable, to evince the existence of the nature we examine. And, though I should admit the use that some divines contend for of the holy scriptures in philosophical controversies, yet I should not be persuaded of the existence of the nature we dispute of. For I do not remember that the scripture anywhere declares to us that there is such a thing (in the sense by me questioned), though (as I formerly noted more fully in the fourth section) in Genesis and some other places where the corporeal works of God are expressly treated of (though in order to spiritual ends), one might probably enough expect to find some mention of God's grand vicegerent in the universe of bodies, if he had established any such. But, whatever be the true cause of the scripture's silence about this matter, the silence itself is sufficient to justify me for examining freely by reason a thing that is not imposed on my belief by revelation. And, as for the physical arguments that may be brought in favour of the questioned notion of nature, I shall ere long examine the principal of them and show that they are not convincing. To these things may be added, as to

[a] The opening paragraph and the first three arguments were written in the 1660s; see BP 8, fols. 149–52, and BP 10, fols. 103–6.

58

the proof drawn from the general opinion about nature: that being a popular, not a physical argument, it may indeed pass for current with the vulgar, but ought not to do so with philosophers.

2. The second reason is taken from the unnecessariness of such a nature, as is pretended. For since a great part of the work of true philosophers has been to reduce the principles of things to the smallest number they can, without making them insufficient, I see not why we should take in a principle of which we have no need. For supposing the common matter of all bodies to have been at first divided into innumerable minute parts by the wise author of nature, and these parts to have been so disposed of as to form the world, constituted as it now is; and especially, supposing that the universal laws of motion among the parts of the matter have been established, and several conventions of particles contrived into the seminal principles of various things; all which may be effected by the mere local motion of matter (not left to itself, but skilfully guided at the beginning of the world) – if (I say) we suppose these things, together with God's ordinary and general concourse, which we very reasonably may, I see not why the same phenomena that we now observe in the world should not be produced, without taking in any such powerful and intelligent being, distinct from God, as nature is represented to be. And till I see some instance produced to the contrary, I am like to continue of this mind and to think that the phenomena we observe will genuinely follow from the mere fabric and constitution of the world. As, supposing the sun and moon to have been put at first into such motions about the earth as experience shows they have, the determinate celerity of these motions and the lines wherein they are performed will make it necessary that the moon should be sometimes full, sometimes scarce illuminated at all to us-ward, sometimes horned, and in a word should exhibit such several phases as every month she does, and that at some times she and the sun should have a trine or a quadrate aspect,[b] etc., and that now one, and now the other of them, should at set times suffer an eclipse: though these eclipses were by the Romans and others of old, and are by many unlearned nations at this day, looked upon as supernatural things, and though also Aristotle and a multitude of his followers fancied that such regular motions could not be maintained without an assistant intelligence, which he and they therefore assigned to each of the heavenly orbs.

[b] Angular relations between planets in the heavens of 120 degrees and 90 degrees respectively.

And indeed, the difficulty we find to conceive how so great a fabric as the world can be preserved in order and kept from running again to a chaos seems to arise from hence: that men do not sufficiently consider the unsearchable wisdom of the divine architect or Δημιοῦργός (as the scripture styles him)[1] of the world, whose piercing eyes were able to look at once quite through the universe and take into his prospect both the beginning and end of time, so that, perfectly foreknowing what would be the consequences of all the possible conjunctures of circumstances into which matter, divided and moved according to such laws, could (in an automaton so constituted as the present world is) happen to be put, there can nothing fall out – unless when a miracle is wrought – that shall be able to alter the course of things, or prejudice the constitution of them, any further than he did from the beginning foresee and think fit to allow.

Nor am I sure that the received notion of nature, though it be not necessary, is at least very useful to explicate physical phenomena. For besides that, I shall show ere long that several explications, where recourse to it is presumed to be the most advantageous, are not to be allowed. To give the nature of a thing for the cause of this or that particular quality or operation of it, is to leave men as ignorant as they were before; or, at least, is to acknowledge that a philosopher can, in such cases, assign no better particular and immediate causes of things than a shepherd or a tradesman that never learned natural philosophy can assign of the same things, and of a thousand others. And, though it be true (as I formerly also intimated) that in many cases philosophers themselves can answer no otherwise to such questions as may be proposed to them, than by having recourse to the nature of the thing, yet such answerers do not declare the proper cause of a dark phenomenon, but only that he who employs them does not yet know it. And so this indefinite notion of nature, which is equally applicable to the resolving of all difficulties, is not useful to disclose the thing, but to delude the maker of the question or hide the ignorance of the answerer.

3.   My third argument is, that the nature I question is so dark and odd a thing, that it is hard to know what to make of it – it being scarce (if at all) intelligibly proposed by them that lay most weight upon it. For it appears not clearly, whether they will have it to be a corporeal

---

[1]   Hebrews 11:10.

substance, or an immaterial one, or some such thing as may seem to be betwixt both, such as many Peripatetics do represent substantial forms and what they call real qualities, which divers schoolmen hold to be (at least by miracle) separable from all matter whatsoever. If it be merely corporeal, I confess I understand not how it can be so wise and almost omniscient an agent, as they would have it pass for. Besides that, if it be a body, I would gladly know what kind of body it is? And how, since among bodies there can be no penetration of dimensions, this body can so intimately pervade, as they pretend nature does, all the other bodies of the world? And to this I would add divers other questions that would not be easily answered. (But I shall resume this third argument in another place.)[c] If it be said that nature is a *semi-substantia*, as some of the modernest schoolmen are pleased to call substantial forms and real qualities, I roundly answer that I acknowledge no such chimerical and unintelligible beings, and shall only desire you to apply to them a good part of the discourse made in certain papers, occasioned by 'A Chemico-Physical Essay about Salt-petre',[d] against the pretended origin and inexplicable nature of the imaginary substantial forms of the Peripatetics.

It remains, therefore, that this nature we speak of, if it be anything positive, should be an immaterial substance. But to have recourse to such an one as a physical agent, and not only a determiner but the grand author of the motion of bodies (and that especially in such familiar phenomena as the ascension of water in pumps, the suspension of it in watering pots for gardens, the running of it through siphons, and I know not how many others), and to explain its causality, as they speak, will (I think) prove a work exceeding difficult; though I shall not here spend time to show you the farther inconveniences of such a supposition, being to do that hereafter, and in the meanwhile, contenting myself to observe as to many of the naturists, that though their doctrine may favour it, they seem rather content to talk darkly and uncertainly of what they call nature, than by clearly defining it, expose it to objections not easy to be answered, and who foresee the advantage that the unsettledness of the notion gives them to pretend knowledge, or

---

[c] See the start of section VIII.
[d] 'A physico-chymical Essay ... of Salt-Petre' was printed as part of *Certain Physiological Essays* (London, 1661). Boyle saw *The Origin of Forms and Qualities* (Oxford, 1666) as stemming from it.

disguise ignorance. If Aristotle had considered nature to be an immaterial substance of this sort, then there would have been no need, in my opinion, to attribute particular intelligences to the motions of the heavenly spheres.[e]

4. Since many of the most learned amongst the naturists are Christians, and not few of them divines too, it may not be improper (which else I should perhaps think it would be) to add in this place, that the next thing for which I dislike the vulgar notion (or idea) of nature is, that I think it dangerous to religion in general and, consequently, to the Christian. For this erroneous conceit defrauds the true God of divers acts of veneration and gratitude that are due to him from men, upon the account of the visible world, and diverts them to that imaginary being they call nature, which has no title to them. For, while nature is supposed to be an intelligent thing that wisely and benignly administers all that is done among bodies, it is no wonder that the generality of philosophers and (after their example) of other men should admire and praise her for the wonderful and for the useful things that they observe in the world. And, in effect, though nature in that sense of the word I am speaking of be never (that I remember) to be found in the sacred writings, yet nothing is to be more frequently met with (and that adorned with titles and encomiums) in the books of philosophers than nature and her effects. And if we consider that, whatever has been said by some in excuse of Aristotle himself, yet the generality of the Peripatetics, from whom the vulgar notion of nature is chiefly received, made the world to be eternal, and referred all the transactions among the bodies it contains to what they called nature. Whence it will not be difficult to perceive that if they do not quite exclude God, yet, as they leave him no interest in the first formation of the universe, so they leave him but very little in the administration of the parts it consists of, especially the sublunary ones. So that instead of the true God, they have substituted for us a kind of a goddess with the title of 'nature', which, as they look upon as the immediate agent and director in all excellent productions, so they ascribe to her the praise and glory of them.

Whether this great error in a point of such importance may not undermine the foundation of religion, I think it may not irrationally be

---

[e]   This sentence was printed only in Lat. which has: *Naturam si Aristoteles immaterialem eiusmodi substantiam existimasset, necesse, meo quidem judicio, non fuisset ut particulares Coelestium Orbium motibus intelligentias præfecisset.*

suspected. For, since the most general and efficacious argument that has persuaded philosophers and other men that there is a God and a providence is afforded by the consideration of the visible world – wherein so many operations and other things are observed that are managed (or performed) with such conduct and benignity as cannot justly be ascribed but to the wisdom and goodness of a deity – they that ascribe these things to mere nature do much weaken the force of that argument, if they do not quite take away the necessity of acknowledging a deity, by showing that, without any need of having recourse to him of the administration of the world and of what is performed among things corporeal, an account may be given. Though, when men are put upon considering the matter and pressed to declare themselves more clearly, they are ashamed to affirm that God and nature are the same thing, and will confess that she is but his vicegerent; yet, in practice, their admiration and their praises are frequently given to nature, not to God: in like manner as, though the sun be the fountain of light, and the moon derives all hers from the sun, yet the sea, in its grand motions of ebbing and flowing, appears to respect the moon and not the sun; for thus the generality of men, though they will acknowledge that nature is inferior and subordinate to God, do yet appear to regard her more than him.

To be short, nature uses to be so frequently recurred to, and is so magnified in the writings of physiologers, that the excessive veneration men have for nature, as it has made some philosophers (as the Epicureans) deny God, so it is to be feared that it makes many forget him. And perhaps a suspicious person would venture to add that, if other principles hindered not (as I know that in many, and think that in most, of the Christian naturists they do), the erroneous idea of nature would too often be found to have a strong tendency to shake, if not to subvert, the very foundations of all religion, misleading those that are inclined to be its enemies from overlooking the necessity of a God, to the questioning, if not to the denial, of his existence.[f]

5. My fifth and last argument is taken from hence: that I observe divers phenomena which do not agree with the notion or representation of nature that I question. For if indeed there were such an intelligent, powerful and vigilant being as philosophers are wont to describe nature to be, divers things would not be done which experience assures us are done.

[f]   Lat. lacks the phrases after 'religion'.

And here I shall once for all give an advertisement, which I desire may be called to mind whenever there shall be occasion, in the following part of this tract, which is this: that, because inanimate bodies are usually more simple, or less compounded, and of a slighter and less complicated or curious contrivance, than animals or plants, I thought fit to choose most of the instances I employ, rather among lifeless bodies, whose structure and qualities are more easy to be intelligibly and with brevity discoursed of, than among living creatures, whose textures, being organical, are much more intricate and subtle. And this course I did not scruple to take, because the celebrators of nature give her a province, or rather an empire, as large as the world, and will have her care and jurisdiction reach as well to inanimate as to living bodies. And accordingly, most of the conspicuous instances they allege of her providence and power are taken from bodies destitute of life, as when they tell us that the ascension of water in sucking pumps and the sustentation of it in gardeners' watering pots are caused by nature's abhorrence of a vacuum; that heavy bodies (unhindered) fall to the ground in a perpendicular line because nature directs them the shortest way to the centre of the earth; and that bubbles rise through the water, and flames ascend in the air, because nature directs these bodies to rejoin themselves to their respective elements; to omit other instances of this sort that there will be occasion to mention hereafter – till when, these may suffice to warrant my taking most of my instances from inanimate bodies, though I shall not confine myself to these, especially when I shall come to answer objections that are taken from living creatures.

The foregoing advertisement will be (I hope) found conducive to clear the way for my fifth argument, lately proposed, which concludes that if indeed there were such a being as nature is usually represented to be, several things would be otherwise administered in the universe than experience shows they are. To enumerate all the particulars that may be proposed to make this good, would swell this discourse much beyond the bulk to which my haste obliges me to confine it. But to make you amends for the paucity of instances I shall now name by the kind of them, I shall propose such as for the most part are taken from those very things whence the wisdom and vigilancy of nature is wont to be confidently argued – which I the rather do, that by such I may make way for, and shorten the answers I am to give to, the arguments ere long to be examined.

First then, whereas the great care and vigilancy of nature for the common good of the universe is wont to be demonstrated from the watchful care she takes to prevent or replenish a vacuum, which would be very prejudicial to the fabric of the world: I argue the quite contrary from the phenomena that occur about a vacuum. For whereas it is alleged that nature, in great pumps and in the like cases, lifts up the heavy body of water in spite of its tendency towards the centre of the earth, to obviate or fill up a vacuity; and that out of a gardener's pot or inverted pipe stopped at one end, neither the water, nor even quick-silver that is near fourteen times as heavy, will fall down, lest it should leave a vacuum behind it: I demand how it comes to pass, that if a glass pipe be but a foot longer than 34 or 35 feet, or an inverted tube filled with quicksilver be but a finger's breadth longer than 30 inches, the water in the one and the quicksilver in the other will subside, though the one will leave but about a foot, and the other but about an inch, of deserted space, which they call vacuum, at the top of the glass. Is it possible that nature, that in pumps is said to raise up every day so many hundred tons of water, and (if you will believe the schools) would raise it to any height (lest there should be a vacuum), should not have the discretion or the power to lift up or sustain as much water as would serve to fill one foot in a glass tube, or as much quicksilver as an inch of a slender pipe will contain, to obviate or replenish the vacuum she is said so much to abhor? Sure, at this rate, she must either have very little power or very little knowledge of the power she has.

So likewise, when a glass bubble is blown very thin at the flame of a lamp and hermetically sealed while it is very hot, the cause that is rendered why it is apt to break when it grows cold, is that the inward air (which was before rarefied by the heat), coming to be condensed by the cold, lest the space deserted by the air that thus contracts itself should be left void, nature with violence breaks the glass in pieces. But, by these learned men's favour, if the glass be blown but a little stronger than ordinary, though at the flame of a lamp, the bubble (as I have often tried) will continue unbroken, in spite of nature's pretended abhorrency of a vacuum, which needs not at all to be recurred to in the case. For the reason why the thin glass bubble broke not when it was hot, and did when it grew cold, is plainly this: that in the former state, the agitation of the included air by the heat did so strengthen the spring of it, that the glass was thereby assisted and enabled to resist the weight of the incumbent air. Whereas, upon the cessation of that heat, the debilitated spring of the internal [air], being unable to assist the glass as formerly to resist the pressure of the external air, the glass itself being too thin becomes unable to support the weight or pressure of the incumbent air, the atmospherical pillar that leans upon a bubble of about two inches diameter, amounting to above one hundred pound weight, as may be manifestly concluded from a late experiment that I have tried, and you may meet with in another paper.[g] And the reason why, if the bubble be blown of a due thickness, it will continue whole after it is cold, is that the thickness of it, though but faintly assisted by the weakened spring of the included air, is sufficient to support the weight of the incumbent air; though several times I have observed the pressure of the atmosphere and the resistance of the bubble to have been, by accident, so near the equipollent that a much less outward force than one would imagine applied to the glass, as perhaps a pound or a less weight gently laid on it, would enable the outward air to break it with noise into a multitude of pieces.

And now give me leave to consider how ill this experiment and the above-mentioned phenomena that happen in glass pipes, wherein water and quicksilver subside, agree with the vulgar apprehension men have of nature. For, if in case she did not hinder the falling down of the water or the quicksilver, there would be no such vacuum produced as she is said to abhor. Why does she seem so solicitous to hinder it? And why does she keep three or four and thirty foot of water in perpendicular height, contrary to the nature of all heavy bodies, suspended in the tube? And why does she furiously break in pieces a thin sealed bubble, such as I come from speaking of, to hinder a vacuum? If in case she did not break it, no vacuum would ensue. And on the other side, if we admit her endeavours to hinder a vacuum not to have been superfluous, and consequently foolish, we must confess that where these endeavours succeed not, there is really produced such a vacuum as she is said to abhor. So that, as I was saying, either she must be very indiscreet to trouble herself and to transgress her own ordinary laws to prevent a danger she need not fear, or her strength must be very small – that is, not able to fill a vacuity that half a pint of water or an ounce of quicksilver may replenish; or break a tender glass bubble, which

[g] Perhaps a reference to *New Experiments Physico-Mechanical, Touching the Spring of the Air*, which contains several experiments on hermetically sealed glasses.

perhaps a pound weight on it would, with the help of so light a body as the incumbent air, crush in pieces.

The other grand instance that is given of the wisdom of nature and her watchfulness for the good of the whole world is the appetite she has implanted in all heavy bodies to descend to the centre of the earth, and in all light ones, to ascend towards heaven – or, as some would have it, towards the element of fire, contiguous to the orb of the moon. But, for positive levity, until I see it better proved than it has hitherto been, I allow no such thing implanted in sublunary bodies, the prepollent gravity of some sufficing to give others a comparative or respective lightness. As a piece of oak or the like wood, being let go in the air, falls down by its own gravity, or rather by virtue of the efficient [cause] of that gravity. But if it be let go under water, it will, though it be never so great a log or piece of timber, ascend with a considerable force to the top of the water – which I hope will not be ascribed to a positive levity, since when it descended in the air, it was by its gravity that it did so. But not to insist on this, nor to take notice how wisely nature has implanted into all heavy bodies an appetite to descend to the centre of the earth, which, being but a point, is not able to contain any one of them.

Not to urge these things (I say), I will only invite you to consider one of the most familiar things that occur among heavy bodies. For if (for example) you let fall a ball upon the ground, it will rebound to a good height, proportionable to that from whence you let it fall, or perhaps will make several lesser rebounds before it come to rest. If it be now asked, why the ball, being let out of your hand, does not fall on this or that side, or move upwards, but falls directly towards the centre of the earth by that shortest line (which mechanicians call *linea directionis* [the line of direction]) which is the diameter of the earth prolonged to the centre of gravity of the ball? It will be readily answered that this proceeds from the ball's gravity, i.e. an innate appetite whereby it tends to the centre of the earth the nearest way. But then I demand, whence comes this rebound, i.e. this motion upwards? For it is plain, it is the genuine consequence of the motion downwards, and therefore is increased according as that motion in the ball was increased, by falling from a greater height. So that it seems that nature does in such cases play a very odd game, since she forces a ball, against the laws of heavy bodies, to ascend divers times upwards, upon the account of that very gravity whose office it is to carry it downwards the directest way. And at least she seems, in spite of the wisdom ascribed to her, to take her measures very ill, in making the ball move downwards with so much violence, as makes it divers times fly back from the place she intended it should go to. As if a ball which a child can play with and direct as he pleases were so unwieldy a thing that nature cannot manage it, without letting it be hurried on with far greater violence than her design requires.

The reflection I have been making on a ball may (*mutatis mutandis* [taking differences into account], as they speak) be applied to a pendulum. For, since it is unanimously affirmed by all that have written of it, that it falls to the perpendicular upon the account of its gravity, it must not be denied that it is from a motion proceeding from the same gravity that the swinging weight passes beyond the perpendicular, and consequently ascends, and oftentimes makes a multitude of diadroms or vibrations, and consequently does very frequently ascend before it comes to rest in the perpendicular, which is the position wherein its gravity is best complied with, and which therefore it had been best settled in at first.

I shall not here mention those grand anomalies or exorbitances even in the vaster bodies of the universe, such as earthquakes that reach some hundreds of leagues, deluges, destructive eruptions of fire, famines of a large spread, raging pestilences, celestial comets, spots in the sun that are recorded to have obscured it for many months; the sudden appearing, the disappearing and the reappearing of stars that have been judged to be as high as the region of the fixed ones. I will not (I say) enquire how far these anomalies agree to the character wont to be given of nature's watchfulness and vigilancy, because probably I may have hererafter a fit opportunity to do it, and must now proceed to the remaining instances I promised you, which are taken from what happens to animals, as soon as I shall have dispatched some considerations and advertisements that seem necessary to be premised to what I have to offer about that difficult subject.[h]

But to return thither whence I began to make this excursion, perhaps, Eleutherius, you will object against the examples I have produced before it, that the exceptions I have taken at some of the proceedings of

[h] Lat. lacks the clause after 'animals'. Immediately after this the first edition has five paragraphs that, according to the 'Advertisement', actually belong later in this section, on pp. 71–3.

nature may be as well urged against providence, and exclude the one as well as the other from the government of the world. But to this I answer, that this objection is foreign to the question, which is about men's notion of nature, not God's providence – which, if it were here my task to assert, I should establish it upon its proper and solid grounds, such as the infinite perfections of the divine nature, which both engage and enable him to administer his dominion over all things; his being the author and supporter of the world; the exquisite contrivance of the bodies of animals, which could not proceed but from a stupendous wisdom; the supernatural revelations and discoveries he has made of himself; and of his particular care of his creatures, by prophecies, apparitions, true miracles and other ways that transcend the power, or overthrow or at least overrule the physical laws of motion in matter – by these (I say) and the like proper means, I would evince divine providence. But being not now obliged to make an attempt, which deserves to be made very solemnly, and not in such haste as I now write in, I shall at present only observe to you that the case is very differing between providence and nature, and therefore there is no necessity that the objections I have made against the latter should hold against the former.

As (to give you a few instances of the disparity) in the first place, it appears not, nor is it likely, that it is the design of providence to hinder those anomalies and defects I have been mentioning; whereas it is said to be the duty and design of nature, and her only task, to keep the universe in order, and procure in all the bodies that compose it that things be carried on in the best and most regular way that may be for their advantage.

Secondly, nature is confessed to be a thing inferior to God, and so but a subordinate agent, and therefore cannot without disparagement to her power or wisdom or vigilancy suffer divers things to be done, which may, without degradation to God, be permitted by him who is not only a self-existent and independent being, but the supreme and absolute Lord, and if I may so speak, the proprietor of the whole creation – whence both Melchizedek and Abram style him (Genesis 14:19 and 23) not only 'the most high God', but קֹנֵה *Koneh*, 'possessing' (or, as our version has it, 'possessor of') 'heaven and earth'[i] – and who, when he made the world and established the laws of motion, gave them to matter,

---

[i] Lat. lacks the phrase between the dashes.

not to himself. And so, being obliged to none, either as his superior or benefactor, he was not bound to make or administer corporeal things after the best manner that he could, for the good of things themselves – among which, those that are capable of gratitude ought to praise and thank him for having vouchsafed them so much as they have, and have no right to except against his having granted them no more. And – as being thus obliged to none of his works, he has a sovereign right to dispose of them – so he has other attributes which he may justly exercise and both intend and expect to be glorified for, besides his goodness to inferior creatures. And his wisdom may be better set off to men, and perhaps to angels or intelligences, by the great variety of his contrivances in his works, than by making them all of the excellentest kind; as shadows in pictures, and discords in music, skilfully placed and ordered, do much recommend the painter and the musician. Perhaps it may be added, that the permitting the course of things to be somewhat violated shows, by the mischief such exorbitances do, how good God has been in settling and preserving the orderly course of things.

Thirdly, as God is a most absolute and free, so he is an omniscient being, and, as by his supreme dominion over the works of his hands he has a right to dispose of them as he thinks best for his own glory, so, upon the force of his unfathomable wisdom, he may have designs and (if I may so speak) reaches in the anomalies that happen in the world, which we men are too short-sighted to discern; and may exercise as much wisdom, nay, and as much providence (in reference to man, the noblest visible object of his providence) in sometimes (as in divine miracles) receding from what men call the laws of nature, as he did at first in establishing them. Whereas the office of nature being but to preserve the universe in general, and particular bodies in it, after the best manner that their respective conditions will permit, we know what it is she aims at, and consequently can better discern when she misses of her aims by not well acting what is presumed to be her part.

Fourthly, we must consider that, as God is an independent, free and wise, so he is also a just agent, and therefore may very well be supposed to cause many irregularities and exorbitances in the world to punish those that men have been guilty of. And, whereas nature is but a nursing mother to the creatures, and looks even upon wicked men not in their moral but in their physical capacities, God expressly declares in the sacred scriptures that upon Adam's fall he 'cursed the ground, or earth, for man's sake' (Genesis 3:17–18), and that there is no penal 'evil in the city' that is not derived from him (Amos 3:6). He is not overruled, as men are fain to say of erring nature, by the headstrong motions of the matter, but sometimes purposely overrules the regular ones to execute his justice. And therefore plagues, earthquakes, inundations and the like destructive calamities, though they are sometimes irregularities in nature, yet for that very reason they are designed by providence, which intends by them to deprive wicked men of that life, or of those blessings of life, whereof their sins have rendered them unworthy. But, while I mention designs, I must not forget that mine was only to give you a taste of the considerations by which one may show that such things as manifest nature to act unsuitably to the representation that is made of her, may yet, when attributed to divine providence, be made out to have nothing inconsistent with it.[j]

If the past discourse give rise to a question – whether the world and the creatures that compose it are as perfect as they could be made? – the question seems to me, because of the ambiguity of the terms, too intricate to be resolved by a single answer. But yet, because the problem is not wont to be discussed and is, in my opinion, of moment in reference to natural theology, I shall venture briefly to intimate some of the thoughts that occurred to me about it, having first declared that I am, with reason, very backward to be positive in a matter of this nature – the extent of the divine power and wisdom being such that its bounds, in case it have any, are not known to me.

This premised, I consider that the sense of the question may be, whether God could make the material world and the corporeal creatures it consists of, better and more perfect than they are, speaking in a general way and absolute sense? Or else, whether the particular kinds or orders of the creatures in the world, could any of them be made more perfect or better than they have been made?

To answer the question in the first-named sense of it, I think it very unsafe to deny that God, who is almighty and omniscient, and an owner of perfections which, for ought we know, are participable in more different manners and degrees than we can comprehend, could not display – if it be not fitter to say adumbrate – them, by creating a work more excellent than this world. And, his immense power and unexhausted wisdom considered, it will not follow either, that because this

---

[j] The next five paragraphs were printed out of order in the first edition.

world of ours is an admirable piece of workmanship, the divine architect could not have bettered it; or, because God himself is able to make a greater masterpiece, this exquisitely contrived system is not admirably excellent.

But the proposed question, in the other sense of it, will require some more words to resolve it. For if we look upon the several species of visible creatures under a more absolute consideration, without respect to the great system of the universe of which they are parts or to the more particular designs of the creator, it seems manifest that many sorts of creatures might have been more perfect than they are, since they want many completing things that others are endowed with – as an oyster that can neither hear, nor see, nor walk, nor swim, nor fly, etc., is not so perfect a creature as an eagle or an elephant, that have both those senses that an oyster wants, and a far more active faculty of changing places. And of this inequality of perfection in creatures of differing kinds, the examples are too obvious to need to be enumerated. But if the question be better proposed, and it be enquired not whether God could have made more perfect creatures than many of those he has made (for that it is plain he could do, because he has done it); but whether the creatures were not so curiously and skilfully made that it was scarce possible they could have been better made, with due regard to all the wise ends he may be supposed to have had in making them, it will be hard to prove a negative answer.

This I shall endeavour to illustrate by a supposition. If one should come into the well-furnished shop of an excellent watchmaker, and should there see a plain watch designed barely to show the hour of the day; another that strikes the hours; a third that is also furnished with an alarm; a fourth that, besides these, shows the month current and the day of it; and lastly, a fifth that, over and above all these, shows the motions of the sun, moon and planets, the tides and other things which may be seen in some curious watches; in this case, I say, the spectator, supposing him judicious, would indeed think one of these watches far more excellent and complete than another, but yet he would conclude each of them to be perfect in its own kind, and the plain watch to answer the artificer's idea and design in making it as well as the more compounded and elaborate one did. The same thing may, in some circumstances, be further illustrated by considering the copy of some excellent writing master. For though there we may find some leaves written in an Italian hand, others in a secretary[k] and, in others, hands of other denominations; though one of these patterns may be much fairer and more curious than another, if they be compared together; yet if we consider their equal conformity to the respective ideas of the author and the suitableness to the design he had of making each copy not as curious, sightly and flourishing as he could, but as conformable to the true idea of the sort of hand he meant to exhibit and the design he had to show the variety, number and justness of his skill by that of the patterns he made complete in the respective kinds, we shall not think that any of them could have been bettered by him. And if he should have made a text hand as fair as a Roman hand by giving it more beauty and ornament, he would not have made it better in its kind, but spoiled it and, by a flourish of his skill, might have given a proof of his want of judgement.

And yet, somewhat further to clear this weighty matter, and particularly some things but briefly hinted in what I have been lately discoursing, I think it fit, before I descend to the particulars that I am to employ against the vulgar notion of nature, to premise somewhat by way of caution, that I may do some right (though I can never do enough) to divine providence, and take care betimes, that no use injurious to it may be made of anything that my argument has obliged me, or will oblige me, to say about that imaginary thing vulgarly called nature, either in this or the sixth section, or any other part of our present *Enquiry*.

I conceive then, that the divine author of things, in making the world and the particular creatures that compose it, has respect to several ends, some of them knowable by us men, and others hid in the abyss of the divine wisdom and counsels. And that of those ends which are either manifest enough to us, or at least discoverable by human sagacity and industry, some of the principal are: the manifestation of the glory of God, the utility of man and the maintenance of the system of the world, under which is comprised the conservation of particular creatures and also the propagation of some kinds of them.

But this general design of God for the welfare of man and other creatures is not (as I conceive) to be understood but with a twofold

---

[k]   'Italian hand' was an italic handwriting developed in Italy; 'secretary' was a distinctive form of handwriting used especially for legal documents in this period. 'Roman hand', mentioned below, was a round and bold hand.

limitation. For first, though men and other animals be furnished with faculties or powers and other requisites to enable them to preserve themselves and procure what is necessary for their own welfare, yet this provision that God has been pleased to make for them with reference to what regularly, or what most usually, happens to beings of that species or sort that they belong to; but not with regard to such things as may happen to them irregularly, contingently and (in comparision of the others) unfrequently. Thus it is in general far better for mankind that women, when they are brought to bed, should have their breasts filled with milk to give suck to the newborn babe, than that they should not; though sometimes, as if the child die in the delivery or presently after, and in some other cases also, the plentiful recourse of milk to the mother's breasts proves troublesome and inconvenient and sometimes also dangerous to her. Thus a head of hair is, for the most part, useful to the person (whether man or woman) that nature has furnished with it, though in some cases (as of consumptions and in a few other circumstances) it happens to be prejudicial to the wearer, and therefore physicians do often with good success prescribe that it be shaven off. Thus the instinct that hens have to hatch their eggs and take care of their young is in general very useful (if not necessary) for the conservation of that species of birds, and yet it sometimes misguides and deludes them, when it makes them take a great deal of pains to brood upon those duck eggs that housewives (having taken away the bird's own eggs) lay in her nest, which makes her very solicitous to hatch and take care of ducklings instead of chickens. Thus it is an institution that ordinarily is profitable for man, that his stomach should nauseate or reject things that have a loathsome taste or smell, because the generality of those things that are provided for his nourishment are well, or at least not ill tasted; and yet, on some occasions of sickness, that disposition of the stomach to refuse or vomit up nauseous purges and other distasteful medicines (as such remedies are usually loathsome enough) proves very prejudicial, by being a great impediment to the recovery of health. And thus (to be short) the passions of the mind, such as fear, joy and grief, are given to man for his good, and when rightly used, are very advantageous, if not absolutely necessary, to him, though when they grow unruly or are ill managed, as it but too often happens, they frequently prove causes of diseases and of great mischief, as well to the passionate man himself as to others.

The second limitation (which has a natural connection with the former) is this: that the omniscient author of things, who in his vast and boundless understanding comprehended at once the whole system of his works and every part of it, did not mainly intend the welfare of such or such particular creatures, but subordinated his care of their preservation and welfare to his care of maintaining the universal system and primitive scheme or contrivance of his works, and especially those catholic rules of motion, and other grand laws, which he at first established among the portions of the mundane matter. So that, when there happens such a concourse of circumstances that particular bodies (fewer or more) must suffer, or else the settled frame or the usual course of things must be altered, or some general law of motion must be hindered from taking place – in such cases, I say, the welfare and interest of man himself (as an animal), and much more that of inferior animals and of other particular creatures, must give way to the care that providence takes of things of a more general and important nature or condition.

Thus (as I formerly noted) God established the lines of motion which the sun and the moon observe, though he foresaw that from thence there would necessarily from time to time ensue eclipses of these luminaries, which he chose rather to permit, than to alter that course which, on several accounts, was the most convenient. Thus a blown bladder or a football, falling from a considerable height upon the ground, rebounds upwards, and so, contrary to the nature of heavy bodies, moves from the centre of the earth, lest the catholic laws of motion, whereby the springiness and reflection of bodies in such circumstances are established, should be violated or entrenched upon. Thus he thought not fit to furnish sheep with paws, or tusks, or swiftness, or animosity, or craft, to defend or preserve themselves from wolves and foxes and other beasts of prey. And tame and fearful birds, such as hens, are so ill provided for defence that they seem designed to be the food of hawks, kites and other rapacious ones. Thus oysters, having neither eyes nor ears, are not near so well provided for as the generality of beasts and birds and even most other fishes. And thus silkworms (to name no other caterpillars) usually (at least in these countries) live not much above half a year, being less furnished with the requisites of longevity than the generality of birds and beasts and fishes.

I have thought fit to lay down the two foregoing limitations, partly because they will be of use to me hereafter, and partly because they contain something that may be added to what has been lately represented on behalf of the divine providence (as it falls under the naturalist's consideration). For by these limitations, we may perceive that it is not just presently to deny or censure the providence of God whenever we see some creatures less completely furnished to maintain themselves, or some cases less provided for than we think they might be, or seeming anomalies permitted, which we look upon as mischievous irregularities. For the welfare of men, or of this or that other particular sort of creatures, being not the only – nor in likelihood, the principal – end of God in making the world, it is neither to be admired nor reprehended that he has not provided for the safety and conveniency of particular beings any further than well consists with the welfare of beings of a more considerable order, and also will comport with his higher ends, and with the maintenance of the more general laws and customs settled by him among things corporeal. So that divers seeming anomalies and incongruities, whence some take occasion to question the administration of things and to deny the agency of providence, do not only comport with it, but serve to accomplish the designs of it.

I have the more expressly declared my mind on this occasion because, indeed, of the two main reasons which put me upon so difficult a work as I foresaw this treatise would be; as one was the love I bear to truth and philosophical freedom, so the other was a just concern for religion. For thinking it very probable that in so inquisitive an age as this, some observations like mine about nature itself might come into the minds of persons ill affected to divine providence, who would be glad and forward to wrest them and make a perverse use of them, I thought it better that such notions should be candidly proposed by one that would take care to accompany them with those cautions that may keep them from being injurious to religion.

Having premised the two foregoing advertisements to obviate misconstructions, I hope I may now safely proceed to particulars – whereof, for brevity's sake, I shall here mention but a few, leaving you to add to them those others that occur in other parts of this treatise. In the first place, then, I shall take notice that there are several instances of persons that have been choked with a hair, which they were unable either to cough up or swallow down. The reason of this fatal accident is probably said to be the irritation that is made by the stay of so unusual a thing as a hair in the throat, which irritation occasions very violent and disorderly or convulsive motions to expel it in the organs of respiration, by which means the continual circulation of the blood, necessary to the life of man, is hindered, the consequence whereof is speedy death. But this agrees very ill with the vulgar supposition of such a kind and provident being, as they represent nature, which is always at hand to preserve the life of animals, and succour them in their (physical) dangers and distresses, as occasion requires. For since a hair is so slender a body that it cannot stop the throat so as to hinder either the free passage of meat and drink into the stomach or that of the air to or from the lungs (as may be argued from divers no-way mortal excrescences and ulcers in the throat) – were it not a great deal better for nature to let the hair alone, and stay until the juices of the body have resolved or consumed it or some favourable accident have removed it, than like a passionate and transported thing, oppose it like a fury with such blind violence as, instead of ejecting the hair, expels the life of him that was troubled with it?

How the care and wisdom of nature will be reconciled to so improper and disorderly a proceeding, I leave her admirers to consider. But it will appear very reconcilable to providence, if we reflect back upon the lately given advertisement. For in the regard of the use and necessity of deglutition, and in many cases of coughing and vomiting, it was in the general most convenient, that the parts that minister to these motions should be irritated by the sudden sense of things that are unusual (though perhaps they would not be otherwise dangerous or offensive), because (as we formerly noted) it was fit that the providence of God should, in making provision for the welfare of animals, have more regard to that which usually and regularly befalls them, than to extraordinary cases or infrequent accidents.

Though most women are offended with the stink of the smoking wick of a candle, which is no more than men also are, yet it has been frequently observed that big-bellied women have been made to miscarry by the smell of an extinguished candle, which would before have indeed displeased but not endangered the same persons. So that it seems nature is, in these cases, very far from being so prudent and careful as men are wont to fancy her, since by an odour (which, if calmly received, would have done no harm to the teeming woman) she is put into such unruly transports. And instead of watching for the welfare of the teeming woman, whose condition needed a more than ordinary measure of her care and tenderness, she violently precipitates her poor charge into a danger that oftentimes proves fatal, not only to the mother, but the child also.

The improper and oftentimes hurtful courses that nature takes in persons that are sick, some of one disease, some of another, will be hereafter taken notice of in opportune places. And therefore for the present I shall only observe that nature seems to do her work very weakly or bunglingly in the production of monsters, whose variety and numerousness is almost as great as their deformity or their irregularity, in so much that several volumes have been written, and many more might have been, to give the description of them.[1] How these gross aberrations will agree with that great uniformity and exquisite skill that is ascribed to nature in her seminal productions, I leave the naturists to make out. I know that some of them lay the fault upon the stubbornness of the matter that would not be obsequious to the plastic power of nature, but I can hardly admit of this account from men of such principles as they are that give it. For it is strange to me they should pretend that nature, which they make a kind of semi-deity, should not be able to mould and fashion so small and soft and tractable a portion of matter as that wherein the first model and efformation of the embryo is made, when, at the same time, they tell us that it is able in sucking pumps to raise and, if need be, sustain whole tons of water, to prevent a vacuum; and can in mines toss up into the air houses, walls, and castles, and (perhaps) the rocks they are built on, to give the kindled gunpowder the expansion that its new state requires.

Other arguments that, by a light change and easy application, may be made use of and added to these against the vulgar notion of nature, may be met with in divers parts of this treatise, and especially in the seventh section, for which reason (among others) I decline lengthening this part of my discourse with the mention of them.[m]

---

[1] For such books, see Katharine Park and Lorraine J. Daston, 'Unnatural Conceptions: The Study of Monsters in Sixteenth and Seventeenth-Century France and England', *Past and Present*, 92 (1981), 20–54.

[m] According to the 'Advertisement', this is where Boyle intended this section to end. What follows in the first edition actually belongs at the end of section VI, where we have placed it.

SECTION VI

Having in the foregoing section proposed some of the considerations that have dissatisfied me with the received notion of nature, it may now be justly expected that I should also consider what I foresee will be alleged in its behalf by the more intelligent of its favourers. And I shall not deny the objections I am going to name against my opinion to be considerable, especially for this reason: that I am very unwilling to seem to put such an affront upon the generality as well of learned men as of others, as to maintain that they have built a notion of so great weight and importance upon slight and inconsiderable grounds. The reasons that I conceive may have induced philosophers to take up and rely on the received notion of nature are such as these that follow.

And the first argument, as one of the most obvious, may be taken from the general belief – or, as men suppose, observation – that divers bodies, as particularly earth, water, and other elements, have each of them its natural place assigned it in the universe; from which place, if any portion of the element, or any mixed body wherein that element predominates, happens to be removed, it has a strong incessant appetite to return to it, because when it is there it ceases either to gravitate or (as some schoolmen speak) to levitate, and is now in a place which nature has qualified to preserve it, according to the axiom that *locus conservat locatum* [a place keeps (or maintains) whatever is located there].

To this argument I answer, that I readily grant that, there being such a quantity of very bulky bodies in the world, it was necessary they should have places adequate to their bigness; and it was thought fit by the wise architect of the universe that they should not be all blended together, but that a great portion of each of them should at the beginning of things be disposed of and lodged in a distinct and convenient place. But when I have granted this, I see not any necessity of granting likewise what is asserted in the argument above proposed. For inanimate bodies having no sense or perception (which is the prerogative of animadversive beings), it must be all one to them in what place they are, because they cannot be concerned to be in one place rather than in another, since such a preference would require a knowledge that inanimate things are destitute of. And for the same reason, a portion of an element, removed by force or chance from what they call its proper place, can have no real appetite to return thither. For who tells it it is in an undue place, and that it may better its condition by removing into another? And who informs it whether that place lies on this hand of it, or that hand of it, or above it, or beneath it? Some philosophers indeed have been somewhat aware of the weakness of the argument drawn from the vulgarly proposed instance (which yet is the best that is wont to be employed) of earthy bodies, which being let fall from the top of an house, or thrown into the air, do of themselves fall in a direct line towards the centre of the earth. And therefore they have strengthened this argument, as far as might be, by pretending that these bodies have not indeed, as former philosophers were wont to think, an appetite to descend to the centre of the earth, but to the great mass of their connatural bodies. I will not therefore accuse these philosophers of the inconsiderate opinion of their predecessors, who would have nature make all heavy things affect to lodge themselves in the centre of the earth, which (as was formerly noted) being but a point, cannot contain any one of them (how little soever it be). But yet the hypothesis of these moderns is liable, though not to that, yet to other weighty objections.

For the first argument I lately employed will hold good against these philosophers too, it not being conceivable how an inanimate body should have an appetite to rejoin homogeneous bodies, neither whose situation nor whose distance from it it does at all know.

Secondly, it does not appear that all bodies have such an appetite (as is presumed) of joining themselves to greater masses of connatural bodies; as, if you file the end of an ingot or bar of silver or of gold, the filings will not stick to their own mass, though it be approached never so near or made to touch them, and much less will they leap to it when it is at a distance from them. The like may be said almost of all consistent bodies we are acquainted with, except the lodestone and iron, and bodies that participate of one of those two.

Thirdly, it is obvious to them that will observe that that which makes lumps of earth or terrestrial matter fall through the air to the earth is some general agent, whatever that be, which, according to the wise disposition of the author of the universe, determines the motion of those bodies we call heavy, by the shortest ways that are permitted them, towards the central part of the terraqueous globe – whether the body put into motion downwards be of the same, or a like, or a quite differing, nature from the greater mass of matter to which, when it is aggregated, it rests there. If, from the side of a ship, you let fall a chip of wood out of your hand when your arm is so stretched out that the perpendicular (or shortest line) between that and the water lies never so little without the ship, that chip will fall into the sea, which is a fluid body and quite of another nature than itself, rather than swerve in the least from the line of direction (as mechanicians call it) to rejoin itself to the great bulk of wood whereof the ship, though never so big, consists. And, on the other side, if a man standing upon the shore just by the sea shall pour out a glass of water, holding the glass just over his feet, that water will fall into the sand, where it will be immediately soaked up and dispersed, rather than deviate a little to join itself to so great a mass of connatural body as the ocean is.

And as to what is generally believed and made part of the argument that I am answering, that water does not weigh in water because it is in its own natural place, and *elementa in proprio loco non gravitant* [elements in their proper place do not have weight]: I deny the matter of fact, and have convinced divers curious persons by experiment[1] that water does gravitate in water as well as out of it, though indeed it does not pregravitate, because it is counterbalanced by an equal weight of collateral water which keeps it from descending.

And lastly, for the maxim that *locus conservat locatum*;[a] besides that it has been prooflessly asserted and therefore, unless it be cautiously explained, I do not think myself bound to admit it – besides this, I say, I think that either the proper place of a body cannot be inferred, as my adversaries would have it, from the natural tendency of a body to it; or else it will not hold true in general that *locus conservat locatum*. As when, for instance, a poor unlucky seaman falls from the mainyard of a ship into the water, does the sea to which he makes such haste preserve him or destroy him? And when in a foul chimney, a lump of soot falls into the hearth and presently burns up there, can we think that the wisdom of nature gave the soot an appetite to hasten to the fire, as a greater bulk of its connatural body, or a place provided by nature for its conservation?

And now [that] I speak of such an innate appetite of conjunction between bodies, I remember what I lately forgot to mention in a fitter place: that bubbles themselves may overthrow the argument I was

---

[1]   See the Appendix to the *Hydrostat. Paradoxes*. [This note, missing in Lat., cites Boyle's own *Hydrostatical Paradoxes* (Oxford, 1666).]

[a]   Lat. lacks the rest of this sentence.

answering. For if a bubble happens to arise from the bottom of a vessel to the upper part of it, we are told that the haste wherewith the air moves through water proceeds from the appetite it has to quit that preternatural place and rejoin the element, or great mass of air detained at the very surface of the water by a very thin skin of that liquor, together with which it constitutes a bubble. Now I demand how it comes to pass, that this appetite of the air – which, when it was at the bottom of the water, and also in its passage upwards, is supposed to have enabled it to ascend with so much eagerness and force as to make its way through all the incumbent water (which possibly was very deep) – should not be able, when the air is arrived at the very top of the water, to break through so thin a membrane of water as usually serves to make a bubble, and which suffices to keep it from the beloved conjunction with the great mass of the external air, especially since they tell us that natural motion grows more quick, the nearer it comes to the end or place of rest, the appetites of bodies increasing with their approaches to the good they aspire to, upon which account falling bodies, as stones, etc., are said (though falsely) to increase their swiftness the nearer they come to the earth. But if, setting aside the imaginary appetite of the air, we attribute the ascension of bubbles to the gravity and pressure upwards of the water, it is easy hydrostatically to explicate why bubbles often move slower when they come near the surface of the water, and why they are detained there; which last phenomenon proceeds from this: that the pressure of the water being there inconsiderable, it is not able to make the air quite surmount the resistance made by the tenacity of the superficial part of the water. And therefore in good spirit of wine, whose tenacity and glutinousness is far less than that of water, bubbles rarely continue upon the surface of the liquor, but are presently broken and vanish.

And, to make this presumed appetite of the smaller portions of the air to unite with the great mass of it appear the less probable, I shall add that I have often observed that water, in that state which is usually called its natural state, is wont to have store of aerial particles mingled with it – notwithstanding the neighbourhood of the external air that is incumbent on the water – as may appear by putting a glass full of water into the receiver of the new pneumatical engine.[b] For the pressure of the external air being by the pump taken off, there will from time to

[b] i.e. the vacuum pump.

time disclose themselves in the water a multitude of bubbles, made by the aerial particles that lay concealed in that liquor. And I have further tried, as I doubt not but some others also have done, that by exactly enclosing in a conveniently shaped glass some water, thus freed from the air, and leaving a little air at the top of the vessel, which was afterwards set by in a quiet place; the corpuscles of that incumbent air did, one after another, insinuate themselves into the water and remained lodged in it – so little appetite has air in general to flee all association with water and make its escape out of that liquor, though when sensible portions of it happen to be underwater, the great inequality in gravity between those two fluids makes the water press up the air. But, though it were easy to give a mechanical account of the phenomena of mingled air and water, yet because it cannot be done in few words I shall not here undertake it, the phenomena themselves being sufficient to render the supposition of my adversaries improbable.

Another argument in favour of the received opinion of nature may be drawn from the strong appetite that bodies have to recover their natural state, when by any means they are put out of it, and thereby forced into a state that is called preternatural: as we see that air being violently compressed in a blown bladder, as soon as the force is removed will return to its first dimensions; and the blade of a sword being bent by being thrust against the floor, as soon as the force ceases restores itself by its innate power to its former straightness; and water, being made hot by the fire, when it is removed thence hastens to recover its former coldness. But though I take this argument to have much more weight in it than the foregoing, because it seems to be grounded upon such real phenomena of nature as those newly recited, yet I do not look upon it as cogent. In answer to it, therefore, I shall represent that it appears by the instances lately mentioned, that the proposers of the argument ground it on the affections of inanimate bodies. Now an inanimate portion of matter being confessedly devoid of knowledge and sense, I see no reason why we should not think it incapable of being concerned to be in one state or constitution, rather than another, since it has no knowledge of that which it is in at present, nor remembrance of that from which it was forced; and consequently no appetite to forsake the former, that it may return to the latter. But every inanimate body (to say nothing now of plants and brute animals, because I want time to launch into an ample discourse), being of itself indifferent to all places and states, continues in that place or state to which the action and resistance of other bodies, and especially contiguous ones, effectually determine it.

As to the instance afforded by water, I consider that before it be asserted that water being heated return of itself to its natural coldness, it were fit that the assertors should determine what degree or measure of coldness is natural to that liquor – and this, if I mistake not, will be no easy task. It is true indeed that, in reference to us men, water is usually cold, because its minute parts are not so briskly agitated as those of the blood and juices that are to be found in our hands or other organs of feeling. But that water is actually cold in reference to frogs and those fishes that live in it, whose blood is cold as to our sense, has not that I know of been proved, nor is easy to be so. And I think it yet more difficult to determine what degree of coldness is natural to water, since this liquor perpetually varies its temperature (as to cold and heat) according to the temper of the contiguous or the neighbouring bodies, especially the ambient air. And therefore the water of an unshaded pond, for instance, though it rests in its proper and natural place, as they speak, yet in autumn, if the weather be fair, the temperature of it will much vary in the compass of the same day, and the liquor will be much hotter at noon than early in the morning or at midnight; though this great diversity be the effect only of a natural agent, the sun, acting according to its regular course. And in the depth of winter, it is generally confessed that water is much colder than in the heat of summer, which seems to be the reason of what is observed by watermen as a wonderful thing – namely, that in rivers, boats equally laden will not sink so deep in winter as in summer, the cold condensing the water, and consequently making it heavier *in specie* [in specific gravity or density] than it is in summer, when the heat of the ambient air makes it more thin. In divers parts of Africa, that temperature is thought natural to the water, because it is that which it usually has, which is far hotter than that which is thought natural to the same liquor in the frigid zone. And I remember on this occasion, what perhaps I have elsewhere mentioned upon another, that the Russian Czar's chief physician informed me,[c] that in some parts of Siberia (one of the more northern

---

[c] Samuel Collins (1619–70), MD, was physician to Czar Alexis Romanov, 1660–9, and author of the posthumously and anonymously published book *The Present State of Russia* (London, 1671), written in the form of a letter to an unnamed friend who has since been identified as Boyle. See L. Loewenson, 'The Works of Robert Boyle and "The Present State of Russia" by Samuel Collins (1671)', *Slavonic and East European Review*, 33 (1955), 470–85.

provinces of that monarch's empire), water is so much more cold – not only than in the torrid zone, but than in England – that two or three foot beneath the surface of the ground, all the year long (even in summer itself) it continues concreted in the form of ice, so intense is the degree of cold that there seems natural to it.

This odd phenomenon much confirms what I lately intimated of the power of contiguous bodies, and especially of the air, to vary the degree of the coldness of water. I particularly mention the air because, as far as I have tried, it has more power to bring water to its own temperature than is commonly supposed. For though, if in summertime a man puts his hand into water that has lain exposed to the sun, he will usually feel it cold, and so conclude it much colder than the ambient air; yet that may often happen upon another account – namely, that the water being many hundred times a more dense fluid than the air, and consisting of particles more apt to insinuate themselves into the pores of skin, a greater part of the agitation of the blood and spirits contained in the hand is communicated to the water, and thereby lost by the fluids that part with it. And the minute particles of the water, which are perhaps more supple and flexible, insinuating themselves into the pores of the skin – which the aerial particles, by reason of their stiffness and perhaps length, cannot do – they come to affect the somewhat more internal parts of the hand, which being much hotter than the cuticula or scarf-skin, makes us feel them very cold; as when a sweating hand is plunged into lukewarm water, the liquor will be judged cold by him who, if his other hand be very cold, will with it feel the same water hot. To confirm which conjecture I shall add that, having sometimes purposely taken a sealed weather-glass whose included liquor was brought to the tempera-ture of the ambient air, and thrust the ball of it under water kept in the same air, there would be discovered no such coldness in the water as one would have expected; the former reason of the sensible cold the hand feels when thrust into that liquor having here no place. To which I shall add, that having for trial's sake made water very cold by dissolving sal-armoniac in it in summertime, it would, after a while, return to its usual degree of warmth.[d] And having made the same experiment in winter, it would return to such a coldness as belonged to it in that season. So that it did not return to any determinate degree of coldness

---

[d] Ammonium chloride ('sal-armoniac') dissolves endothermically in water, chilling it quite noticeably.

as natural to it, but to that greater or lesser that had been accidentally given it by the ambient air before the sal-armoniac had refrigerated it.

As to the motion of the restitution observable upon the removal or ceasing of the force in air violently compressed and in the blade of a sword forcibly bent, I confess it seems to me a very difficult thing to assign the true mechanical cause of it. But yet I think it far more likely that the cause should be mechanical than that the effect proceeds from such a watchfulness of nature as is pretended. For first I question whether we have any air here below that is in other than a preternatural or violent state, the lower parts of our atmospherical air being constantly compressed by the weight of the upper parts of the same air that lean upon them. As for the restitution of the bent blade of a sword and such like springy bodies when the force that bent them is removed, my thoughts about the theory of springiness belong to another paper.[e] And therefore I shall here, only by way of argument *ad hominem*, consider in answer to the objection that if, for example, you take a somewhat long and narrow plate of silver that has not been hammered or compressed, or (which is surer) has been made red-hot in the fire and suffered to cool leisurely, you may bend it which way you will, and it will constantly retain the last curve figure that you gave it. But if, having again straightened this plate, you give it some smart strokes of a hammer, it will by that merely mechanical change become a springy body: so that if with your hand you force it a little from its rectitude, as soon as you remove your hand it will endeavour to regain its former straightness. The like may be observed in copper, but nothing near so much, or scarce at all, in lead. Now upon these phenomena I demand why, if nature be so careful to restore bodies to their former state, she does not restore the silver blade or plate to its rectitude when it is bent this way or that way before it be hammered? And why a few strokes of a hammer (which, acting violently, seems likely to have put the metal into a preternatural state) should entitle the blade to nature's peculiar care, and make her solicitous to restore it to its rectitude when it is forced from it? And why, if the springy plate be again ignited and refrigerated of itself, nature abandons her former care of it, and suffers it quietly to continue in what crooked posture one pleases to put it into? Not now to demand a

---

[e]   Perhaps a reference to ch. 2 of *An Examen of Mr. T[homas]. Hobbes his Dialogus Physicus*, published in conjunction with the second edition of *New Experiments* (Oxford, 1662); *Works*, vol. 1, pp. 190–7.

reason of nature's greater partiality to silver and copper and iron than to lead and gold itself, in reference to the motion of restitution, I shall add to what I was just now saying, that even in sword blades it has been often observed that though if, soon after they are bent, the force that bent them be withdrawn, they will nimbly return to their former straightness. Yet if they (which are not the only springy bodies of which this has been observed) be kept too long bent, they will lose the power of recovering their former straightness and continue in that crooked posture, though the force that put them into it cease to act. So that it seems nature easily forgets the care she was presumed to take of it at first.

There is an axiom that passes for current among learned men – viz., *nullum violentum durabile* [nothing violent is lasting] – that seems much to favour the opinion of the naturists, since it is grounded upon a supposition that what is violent is, as such, contrary to nature, and for that reason cannot last long. And this trite sentence is by the schools and even some modern philosophers so particularly applied to local motion, that some of them have, not improbably, made it the characteristic token whereby to distinguish natural motions from those that are not so: that the former are perpetual, or at least very durable, whereas the latter, being continually checked more and more by the renitency of nature, do continually decay, and within no long time are suppressed or extinguished. But on this occasion I must crave leave to make the following reflections.

1. It may be justly questioned, upon grounds laid down in another part of this essay, whether there be any motion among inanimate bodies that deserves to be called violent in contradistinction to natural, since among such, all motions where no intelligent spirit intervenes are made according to catholic and (almost, if not more than almost) mechanical laws.

2. Methinks the Peripatetics, who are wont to be the most forward to employ this axiom, should find but little reason to do so, if they consider how unsuitable it is to their doctrine that the vast body of the firmament and all the planetary orbs are, by the *Primum Mobile* [the Prime Mover], with a stupendous swiftness whirled about from east to west in four and twenty hours, contrary to their natural tendency; and that this violent and rapid motion of the incomparably great part of the universe has lasted as long as the world itself – that is, according to Aristotle, for innumerable ages.

3. We may observe here below that the ebbing and flowing of the sea, which is generally supposed to proceed either from the motion of the moon, or that of the terrestrial globe, or some other external cause, has lasted for some thousands of years, and probably will do so as long as the present system of our vortex shall continue. I consider also that the other great ocean, the atmosphere, consists of numberless myriads of corpuscles that are here below continually kept in a violent state, since they are elastical bodies, whereof the lower are still compressed by the weight of the higher. And to make a spring of a body, it is requisite that it be forcibly bent or stretched, and have such a perpetual endeavour to fly open or to shrink in that it will not fail to do so as soon as the external force that hindered it is removed. And as for the states of inanimate bodies, I do not see that their being or not being natural can be with any certainty concluded from their being or not being very durable. For – not to mention that leaves that wither in a few months, and even blossoms that often fade and fall off in [a] few days, are as well natural bodies as the solid and durable trees that bear them – it is obvious that, whether we make the state of fluidity, or that of congelation, to be that which is natural to water, and the other that which is violent; its change from one of those states into another, and even its return to its former state, is oftentimes at some seasons and in some places made very speedily, perhaps in an hour or less, by causes that are acknowledged to be natural. And mists, hail, whirlwinds, lightning, falling stars, to name no more, notwithstanding their being natural bodies, are far from being lasting, especially in comparison of glass, wherein the ingredients, sand, and fixed salt, are brought together by great violence of fire. And the motion that a thin plate or slender wire of this glass can exercise to restore itself to its former position, when forcibly bent, is (in great part) a lasting effect of the same violence of the fire. And so is the most durable perseverance of the indissolubleness of the alcalisate salt that is one of the two ingredients of glass, notwithstanding its being very easily dissoluble in water and other liquors, and not uneasily even in the moist air itself.

There is a distinction of local motion into natural and violent that is so generally received and used, both by philosophers and physicians, that (I think) it deserves to have special notice taken of it in this section, since it implicitly contains an argument for the existence of the thing called nature, by supposing it so manifest a thing as that an important distinction may justly be grounded on it. This implied objection, I confess, is somewhat difficult to clear, not for any great force that is contained in it, but because of the ambiguity of the terms wherein the distinction is wont to be employed, for most men speak of the proposed distinction of motion in so obscure or so uncertain a way, that it is not easy to know what they mean by either of the members of it. But yet some there are who endeavour to speak intelligibly (and for that are to be commended) and define natural motion to be that whose principle is within the moving body itself, and violent motion that which bodies are put into by an external agent or cause. And in regard these speak more clearly than the rest, I shall here principally consider the lately mentioned distinction in the sense they give it.

I say then that, even according to this explication, I am not satisfied with the distinction. For whereas it is a principle received and frequently employed by Aristotle and his followers, *quicquid movetur ab alio movetur* [everything that is moved, is moved by another], it seems that according to this axiom all motion may be called violent, since it proceeds from an external agent. And indeed, according to the school philosophers, the motion of far the greatest part of the visible world, though this motion be most regular and lasting, must, according to the proposed distinction, be reputed violent, since they assert that the immense firmament itself and all the planetary orbs (in comparison of which vast celestial part of the world, the sublunary part is little more than a physical point) is perpetually (and against its native tendency) hurried about the centre of the world once in twenty-four hours, by an external, though invisible, agent, which they therefore call the *Primum Mobile*.

And as for the criterion of natural motion, that its principle is within the moving body, it may be said that all bodies, once in the state of actual motion, whatever cause first brought them to it, are moved by an internal principle; as, for instance, an arrow that actually flies in the air towards a mark moves by some principle or other residing within itself, for it does not depend on the bow it was shot out of, since it would continue though that were broken or even annihilated. Nor does it depend upon the medium, which more resists than assists its progress, as might be easily shown if it were needful. And if we should suppose the ambient air either to be annihilated or (which in our case would be equivalent) rendered incapable of either furthering or hindering its progress, I see not why the motion of the arrow must necessarily cease, since in this case there remains no medium to be penetrated and on that account oppose its progress. When in a watch that is wound up, the spring endeavours to unbend or display itself, and when the string of a drawn bow is broken or let go, the spring of the former and the woody part of the latter does each return to a less crooked line. And though these motions be occasioned by the forcible acts of external agents, yet the watch-spring, and the bow have in themselves (for ought appears to those I reason with) an inward principle by which they are moved till they have attained their position. Some perhaps would add that a squib or a rocket, though an artificial body, seems, as well as a falling star, to move from an internal principle. But I shall rather observe that, on the other side, external agents are requisite to many motions that are acknowledged to be natural, as – to omit the germination and flourishing of divers plants (as onions, leeks, potatoes, etc.), though hung up in the air by the heat of the sun in the spring – to pass by this, I say, if in the pneumatical engine or air pump, you place divers insects (as bees, flies, caterpillars, etc.) and withdraw the common air from the receiver, they will lie moveless, as if they were dead, though it be for several hours while they are kept from enjoying the presence of the air. But when the external air is permitted again to return upon them, they will presently be revived (as I have with pleasure tried), and be brought to move again, according to their respective kinds; as if a fly, for instance, resembled a little windmill in this: that being moveless of itself, it required the action of the air to put its wings and other parts into motion.

But to insist no farther on these arguments *ad hominem*, we may consider that, since motion does not essentially belong to matter, as divisibility and impenetrableness are believed to do, the motions of all bodies, at least at the beginning of things, and the motions of most bodies the causes of whose motions we can discern, were impressed on them either by an external immaterial agent, God, or by other portions of matter (which are also extrinsical impellers) acting on them.

And this occasion invites me to observe that, though motion be deservedly made one of the principal parts of Aristotle's definition of nature,[2] yet men are wont to call such motions natural as are very hard

---

2   *Natura est Principium quoddam et Causa, cur id moveatur et quiescat, in quo inest, etc.* [Nature is the principle and cause of motion and rest in the thing to which it belongs.] *De physica*, ii. 192b 20–2. [The first edition contains a wholly erroneous reference to a spurious work, *Mirabiles auscultationes*.]

to distinguish from those they call violent. Thus, when water falls down to the ground, they tell us that this motion is natural to that liquor, as it is a heavy body. But when a man spurts up water out of his mouth into the air, they pronounce that motion, because of its tendency upwards, to be contrary to nature. And yet when he draws water into his mouth by sucking it through a long pipe held perpendicularly, they will have this motion of the water, though directly upwards, to be not violent, but natural. So when a football or blown bladder, being let fall upon a hard floor, rebounds up to a good height, the descent and ascent are both said to be natural motions, though the former tends towards the centre of the earth and the latter recedes as far as it can do from it. And so, if from a considerable height you let fall a ball of some close wood that yet is not too heavy (as oak or the like) into a deep vessel of water, it will descend a great way in that liquor by a natural motion; and yet its contrary motion upwards ought not to be esteemed violent, since according to the schools, being lighter *in specie* [in specific gravity or density] than water, it is natural to it to affect its proper place, for which purpose it must ascend to the top of the liquor and lie afloat there. And yet it is from these tendencies to opposite points (as the zenith and the nadir) that men are wont to judge many motions of bodies to be natural or violent.

And indeed, since it must be indifferent to a lifeless and insensible body, to what place it is made to move, all its motions may in some respect be said to be natural, and in another, violent. For as very many bodies of visible bulk are set a-moving by external impellents, and on that score their motions may be said to be violent, so the generality of impelled bodies do move either upwards, downwards, etc., towards any part of the world, in what line or way soever they find their motion least resisted; which impulse and tendency, being given by virtue of what they call the general laws of nature, the motion may be said to be natural.

I might here take notice that, according to the Epicurean hypothesis, it need not at all be admitted that motion must be produced by such a principle as the schoolmen's nature. For, according to that great and ancient sect of philosophers, the atomists, every indivisible corpuscle has actual motion, or an incessant endeavour to change places, essentially belonging to it, as it is an atom, in so much that in no case it can be deprived of this property or power. And all sensible bodies being, according to these physiologers, but casual concretions or coalitions of atoms, each of them needs no other principle of motion than that unlosable endeavour of the atoms that compose it; and happen, on the account of circumstances, to have the tendency of the more numerous, or at least the predominant, corpuscles determined one way.

And to these I might add some other such reflections. But I shall in this place say no more concerning motion, not only because – even after having considered the differing definitions that Aristotle, Cartesius and some other philosophers have given of it – I take it to be too difficult a subject to be clearly explicated in few words, but because the only occasion I had to mention it here, was to show that the vulgar distinction of it into natural and violent is not so clear and well grounded as to oblige us to admit (what it supposes) that there is such a being as the naturists assert.

I come now to consider the argument that may be drawn in favour of the received notion of nature from the critical evacuations which happen at certain times in diseases, and the strange shifts that nature sometimes makes use of in them to free herself from the noxious humours that oppressed her.[f] This argument I willingly acknowledge to be very considerable. For we really see that in continual fevers, especially in hotter climates, there do usually happen at certain times of the diseases, notable and critical commotions or conflicts, after which the morbific matter is disposed of and discharged by ways strange and surprising, to the great and speedy relief of the patient, if not to his perfect cure; as may appear by many instances to be met with in the observations of physicians about fevers, pleurisies, etc. Upon this account, I take the argument drawn from crises to be much the weightiest that can be urged for the opinion from which I dissent, and therefore I shall employ the more words in clearing this important difficulty.

In order to this, I desire it may be kept in mind that I do not only acknowledge, but teach, that the body of a man is an incomparable engine, which the most wise author of things has so skilfully framed for lasting very many years, that if there were in it an intelligent principle of self-preservation (as the naturists suppose there is) things would not in most cases be better or otherwise managed for the conservation of the

---

[f]   This paragraph and the next four paragraphs were written *c.* 1680; see BP 18, fols. 103–4; BP 46, fols. 3–5; and BP 2, fol. 190.

animal's life than they generally are. So that the question is not, whether there is a great deal of providence and wisdom exercised in the crises of diseases, but upon what account it is that these apposite things are performed? The universal opinion of physicians is, that it is that intelligent principle they call nature which, being solicitous for the welfare of the patient and distressed by the quantity or hurtfulness of the morbific matter, watches her opportunity (especially when it is concocted) to expel it hastily out of the body, by the most safe and convenient ways which, in the present condition of the patient, can be taken. And I, on the other side, attribute crises to the wisdom and ordinary providence of God, exerting itself by the mechanism, partly of that great machine the world, and partly of that smaller engine the human body, as it is constituted in the patient's present circumstance.

And the reasons that hinder me from acquiescing in the general opinion of physicians about crises are principally these. First, I observe that crises, properly so called, do very seldom happen in other than fevers and the like acute diseases where, according to the common course of things, the malady is terminated in no long time either by recovery, or death, or a change into some other disease. But chronical sicknesses, such as coughs, dropsies, gouts, etc., unless they happen to be accompanied with feverish distempers, are not wont to have crises; which argues that nature does not make critical evacuations upon the account of such care and watchfulness as physicians ascribe them to, since she neglects to employ so salutary an expedient in diseases that are oftentimes no less dangerous and mortal than divers acute diseases which she attempts to cure by crises.

Next, I consider that critical evacuations may be procured by the bare mechanism of the body. For by virtue of that it will often happen that the fibres, or motive organs of the stomach, bowels and other parts, being distended or vellicated by the plenty or acrimony of the peccant matter, will by that irritation be brought to contract themselves vigorously and to throw out the matter that offends the parts, either by the emunctories or common shores of the body, or by whatever passages the proscribed matter can be with most ease discharged. Thus when some men find their stomachs burdened with a clog of meat or drink, they use to thrust their fingers into their throats, and by that mechanical way provoke the stomach to disburden itself of its offensive load, without being beholden to nature's watchfulness for a crisis, which probably she would not (at least so seasonably) attempt. And thus, whereas it is usual enough for crises to be made in fevers by large haemorrhages at the nose, and sometimes at other parts – which is ascribed to nature's watchful solicitude for the patient's recovery – I must take leave to add that it has been divers times observed that, even after death, large bleedings have succeeded at the nose and other parts of the body, which shows that such excretions may be made by virtue of the structure of it and the turgescence and acrimony of the humours, without any design of nature to save the life of the patient, already dead.

Indeed, if it did appear by experience that all, or almost all, the crises of diseases did either expel the morbific matter or at least notably relieve the patient, the critical attempts of nature would much favour the opinion men have conceived of her vigilance and conduct. But unwelcome instances daily show that, as some crises are salutary (as they call them), so others prove mortal. And among those that do not directly or presently kill the patient, there are divers that leave him in a worse condition than he was before. And therefore I wonder not that physicians have thought themselves obliged to lay down several circumstances as necessary requisites of a laudable crisis, if any of which be wanting, it is not thought of the best kind; and if the contrary to some of them happen, it is to be judged either pernicious or at least hurtful. For whereas there are two general ways supposed to be employed by nature in making crises, the one by expulsion of the peccant matter out of the body, and the other by the settling of the matter somewhere within it, neither of these two ways is constantly successful.

And therefore experience has obliged physicians to divide crises not only into perfect, that fully determine the event of the disease, and imperfect, that do but alter it for the better or the worse; but into salutary, that quite deliver the patient, and mortal, that destroy him. And to a perfect and salutary crisis, some learned men require no less than six conditions – namely, that it be preceded by signs of coction of the peccant matter; that it be made by a manifest and sufficiently copious excretion or translation; that it be made upon a critical day, as the seventh, fourteenth, twentieth, etc.; that it leave no relics behind it that may endanger a relapse; that it be made safely, that is, without dangerous symptoms; and lastly, that it be suitable to the nature of the disease and the patient. By this it may appear, that it is no common thing to meet with a perfect and salutary crisis; so many laudable conditions must concur in it. And indeed nature does usually take up with but imperfectly good ones, and it were happy if she made not better, provided she made no worse. But it is found by sad experience that she rouses herself up to make a crisis, not only upon improper and (as physicians call them) intercident days, such as the third, fifth, ninth, etc., or upon those they call empty or medicinal days, which seldom afford any crisis, and much seldomer a good one; but also when there appear not any signs of coction, or at least of due coction, and by these unseasonable attempts weaken the patient and increase the malady, or perhaps make it speedily mortal. Nor will it justify nature to say with some learned physicians, that these attempts are accidentally brought on by the acrimony or importunity of the morbific matter, by which she is provoked before the time to endeavour an expulsion of it. For if nature be indeed so prudent and watchful a guardian as she is thought, she ought not to suffer herself to be provoked to act preposterously and make furious attempts that lavish to no purpose – or worse than no purpose – that little strength the patient has so much need of. And therefore physicians do oftentimes very well when, to act agreeably to the dictates of prudence, they forget how much wisdom they are wont to ascribe to nature and employ their best skill and remedies to suppress or moderate the inordinate motions, or the improper and profuse evacuations, that irritated nature rashly begins to make. And though the crises that are made by a metastasis of the peccant matter, or by lodging it in some particular part of the body, whether external or internal, be oftentimes, when they are not salutary, somewhat less hurtful than those that are made by excretion; yet these do frequently, though perhaps more slowly, prove dangerous enough, producing sometimes inward imposthumes, and sometimes external tumours – in parts that are either noble by their functions, or by their situation, or connection or sympathy with others – that are not to be without hazard or great inconvenience oppressed.

I know that physicians make it a great argument of nature's providence and skill, that she watches for the concoction of the peccant matter before she rouses herself up to expel it by a crisis. What is to be meant by this coction of humours (for it ought not to be confounded with the coction of the aliments) they are not wont so clearly to declare. But as I understand it, when they say that a portion of peccant matter is brought to coction, they mean that it has acquired such a disposition as makes it more fit than before to be separated from the sounder portion of the mass of blood, or from the consistent parts to which it perhaps formerly adhered, and to be afterwards expelled out of the body. This may be partly exemplified by what happens in some recent colds where the lungs are affected, in which we see that after a few days the phlegm is made more fluid, and that which is lodged in the lungs (not sticking so fast to the inside of the *aspera arteria*) is easily brought up by coughing, which could not dislodge it before. And in fevers, that separation in the urine, formerly cloudless, that physicians look upon as a good sign of coction, seems to be produced by some part of the peccant matter that, beginning to be separated from the blood, mingles with the urine and is not usually distinguished from it while this liquor is warm; but when it is grown cold, does on the score of its weight or texture somewhat recede, and appear in a distinct form, as of a cloud, a sediment, etc.[g] But whatever they mean by 'coction', it is plain enough by what has been lately noted, that on many occasions nature does not wait for it, but unseasonably – and oftentimes dangerously – attempts to proscribe the matter that offends her, before it be duly prepared for expulsion.

I come now to that circumstance of crises that is thought the most wonderful, which is that nature does oftentimes by very unusual ways, and at unexpected places, discharge the matter that offends her, and thereby either cures or notably relieves the patient.[h] And it must not be denied that in some cases the critical evacuations have somewhat of surprising in them. And I shall also readily grant that – N.B. – divine providence may expressly interpose, not only in the infliction of diseases by way of punishment, but in the removal of them in the way of mercy. But setting aside these extraordinary cases, I think it not absurd to conjecture that the performances of nature in common crises may be probably referred partly to the particular condition of the matter to be expelled, and partly (and indeed principally) to some peculiar disposition in the primitive fabric of some parts of the patient's body, or some unusual change made in the construction of these parts by the disease itself, or other accidents – [to] which original or adventitious disposition of the sick man's body, not being visible to us, at least while he is alive, we are apt to ascribe the unexpected accidents of a crisis, if it prove

---

[g]   Lat. lacks this sentence.
[h]   This paragraph and the next two paragraphs were written *c.* 1680; see BP 18, fols. 110–11.

salutary to the wonderful providence of nature. And if it happen to be other than salutary, we are wont to overlook them. To illustrate this matter we may consider that plentiful evacuations, procured by medicines, are a kind of artificial crises. We see that some bodies are so constituted that, although the peccant humour wrought on by the medicine ought (as the physician thinks) to be expelled by siege, and indeed is wont to be so in the generality of those that take that kind of medicine (as, for instance, rhubarb or senna), yet the peculiar disposition of the patient's stomach will make that an emetic, which was intended to be, and regularly should be, a cathartic. Nor does this constitution of the stomach equally regard all purging medicines, for the same stomach that will reject them in the form (for instance) of a potion, will quietly entertain them, being in the form of pills.

And to this let me add what we observe of the operation of mercury – which, though if it be duly prepared, it is usually given to procure salivation, especially to succulent bodies; yet there are some patients wherein, instead of salivating, it will violently and dangerously work downwards like a purge, or make some other unexpected evacuation. And I have seen a patient who, though young and very fat, could not be brought to salivate, neither by the gentler ways, nor by turbith-mineral and other harsher medicines, though administered by very skilful physicians and surgeons. And this peculiarity may be as well contracted, as native. For some persons, especially after surfeits, having been roughly dealt with or at length[i] tired out with a medicine of this or that kind of form, will afterwards nauseate and vomit up the like medicine, though in other bodies it be never so far from being emetic. We see also that sometimes sudorific medicines, instead of procuring sweat, prove briskly diuretic, and sometimes either purging or vomitive. From all this we may argue that the qualities of the irritating matter, and much more the particular disposition of the patient's body, may procure evacuations at unexpected places. I remember, too, that among the observations I have met with of famous physicians, there are instances of periodical and critical evacuations at very inconvenient, as well as unusual, vents – as some women are recorded to have had their menses sometimes at the eyes, sometimes at the navel and sometimes at the mouth; of which there seems no cause so probable as some peculiar structure, whether native or adventitious, of the internal parts con-

---

i   Here we follow the MS draft (BP 18, fol. 110v); the printed edition has 'at least'.

cerned in that discharge. And of such unusual structures anatomists must have seen many, since I myself have observed more than one or two.

If these uncommon ways of disposing of the morbific matter were always salutary to the patient, the argument grounded on them would have more weight. But though most men take notice of this sort of crises but when they are lucky, yet an impartial observer shall often find that ill conditioned and hurtful crises may be made by unusual and unexpected ways. And in some translations of the morbific matter to distant and nobler parts, perhaps it will be as difficult to show by what channels or known ways the matter passed from one to another, as it is to determine how it was conducted to those parts at which it was the most happily vented.

In the foregoing discourse about crises, there is (I confess) much of paradox, and it was unwillingly enough that I made an excursion or inroad into a subject that has been looked upon as the physician's peculiar province. And you may remember that, not far from the beginning of this little book, I told you that I was willing to decline meddling with other than inanimate bodies – living ones being as of a less simple sort, so of a more intricate speculation – which reflection will (I hope) excuse me to you, if you find that my proposed brevity, or the difficulty of the subject, has had any great influence on what I write about health, diseases and crises. And as for the sons of Æsculapius,[j] it may be represented to them in my favour that, besides that I have treated of sickness and crises rather as a physiologer than a physician, I could not leave them unconsidered without being thought, if not to betray, at least to be wanting to, the cause I was to plead for.

If it should be disliked that I make the phenomena of the merely corporeal part of the world – under which I comprise the bodies of animals, though not the rational souls of men – to be too generally referred to laws mechanical, I hope you will remember for me several things dispersed in this treatise that may, when laid together, afford a sufficient answer to this surmise. And particularly, that almost all the modern philosophers, and among them divers eminent divines, scruple not to forsake the spread opinion that the celestial orbs were moved and guided by intelligences; and to explicate by physical causes the eclipses

---

[j]   i.e. the physicians: Aesclepius was the Greek god of medicine, to whom physicians swore the oath of Hippocrates.

of the sun and moon, the production or apparition and phenomena of comets and other things that the Romans – as well as other heathens, both ancient and modern – have ascribed to the immediate agency of divine causes. This allows me to observe to you that, since these modern naturalists and divines are wont to explicate the phenomena of the vast celestial bodies by their local motions and the consequences of them, they do, as well as I, endeavour to account for what happens in the incomparably greatest part of the universe, by physico-mechanical principles and laws. And even in the terrestrial part of the world which we men inhabit, most of the moderns that have freed themselves from the prejudices of the schools do not stick to give statical, hydrostatical and other mechanical explications of the ascension of water in pumps, the detention of it in watering pots whose upper orifices are closed, and of other various phenomena which were formerly unanimously ascribed to nature's wonderful providence, expressed in her care to hinder a vacuum.

But perhaps you will think it fitter for me to provide against their censure, who will dislike what I have written about crises not because I have ascribed too much to merely physical causes, but (on the contrary) because I do not strictly confine myself to them. For I doubt that, if you should show these papers to some of your friends that affect to be strict naturalists, they will think it strange that in one of the clauses in the foregoing discourse about crises (I mean that to which this mark 'N.B.' is prefixed),[k] I admit that their events may sometimes be varied by some peculiar interposition of God. But yet I own to you, that the clause it is like they would take exceptions at did not unawares slip from my pen. For it is my settled opinion that divine prudence is often, at least, conversant in a peculiar manner about the actions of men and the things that happen to them, or have a necessary connection with the one, or the other, or both. And though I think it probable that in the conduct of that far greatest part of the universe which is merely corporeal, the wise author of it does seldom manifestly procure a recession from the settled course of the universe, and especially from the most catholic laws of motion; yet where men – who are creatures that he is pleased to endow with free wills (at least in reference to things not spiritual) – are nearly and highly concerned, I think he has, not only sometimes by those signal and manifest interpositions we call miracles, acted by a super-

[k]   See p. 96 l. 25.

natural way, but as the sovereign lord and governor of the world does divers times (and perhaps oftener than mere philosophers imagine) give by the intervention of rational minds – as well united, as not united, to human bodies – divers such determinations to the motion of parts in those bodies, and of others which may be affected by them, as by laws merely mechanical those parts of matter would not have had: by which motions so determined, either salutary or fatal crises, and many other things conducive to the welfare or detriment of men, are produced.[1]

The interposition of divine providences in cases of life and death might be easily shown to Christians out of divers passages of scripture, which expressly proposed long life as a reward to obedient children[3] and to other righteous persons among the Jews, and threatens 'bloody and deceitful men'[4] that they 'shall not live out half their days';[5] and which relates that a king of Israel had his disease made mortal by his impious recourse to the false God of Eckron;[6] and that, upon Hezekiah's prayers and tears, God was pleased to add fifteen years to his life,[7] and grant a special benediction to an outward medicine applied to his threatening sore. To which passages divers may be added out of the New Testament also, and especially that of St James,[8] who exhorts the sick to seek for recovery by prayer; and that of St Paul where, speaking to the Corinthians of the unworthy receivers of the sacrament of the Eucharist, he tells them that, 'For that cause, divers were become sick and weak among them, and many also died.'[9] But though the nature of this discourse dissuades me from employing here the authority of scripture, yet it allows me to observe (what is considerable on this occasion) that natural theology and right reason comport very well with our proposed doctrine. For as I lately intimated and do more fully show in another paper,[10] God has left to the will of man the direction of many local motions in the parts of his own body, and thereby of some others, though the mechanical laws on which the ordinary course of things mainly depends do not only regulate the motions of bodies, but the determinations too. And since man himself is vouchsafed a power to

---

[3] The fifth commandment, in Exodus 20 [verse 12].
[4] Psalms 5:6.   [5] Psalms 55:23.   [6] 2 Kings 1:16.
[7] Isaiah 38.   [8] James 5:25.   [9] 1 Corinthians 11:30.
[10] A Discourse relating to Miracles. [For Boyle's writings on this subject, none published in his own time, see esp. R. L. Colie, 'Spinoza in England 1665–1730', *Proceedings of the American Philosophical Society*, 107 (1963), 183–219.]

[1] Lat. lacks the phrase after the colon.

alter in several cases the usual course of things, it should not seem incredible that the latent interposition of men, or perhaps angels or other causes unthought of by us, should sometimes be employed to the like purposes by God, who is not only the all-wise maker, but the absolute and yet most just and benign rector of the universe and of men.

To conclude the excursion (which I hope will not appear useless) that has been occasioned by the discourse of crises, I think it becomes a Christian philosopher to admit in general that God does sometimes in a peculiar, though hidden way, interpose in the ordinary phenomena and events of crises; but yet that this is done so seldom, at least in a way that we can certainly discern, that we are not hastily to have recourse to an extraordinary providence – and much less to the strange care and skill of that questioned being called nature – in this or that particular case, though perhaps unexpected, if it may be probably accounted for by mechanical laws and the ordinary course of things.

And here, though in a place less proper than I might have chosen if I had timely remembered it, I shall – both in reference to the extraordinary accidents that sometimes happen in crises and more generally to the seemingly irregular phenomena of the universe – venture to offer you a notion that perhaps you will not dislike. I think then that, when we consider the world and the physical changes that happen in it with reference to the divine wisdom and providence, the arguments for the affirmative ought, in their kind, to have more force than those for the negative. For it seems more allowable to argue a providence from the exquisite structure and symmetry of the mundane bodies, and the apt subordination and train of causes, than to infer from some physical anomalies that things are not framed and administered by a wise author and rector. For the characters and impressions of wisdom that are conspicuous in the curious fabric and orderly train of things can with no probability be referred to blind chance, but must be [ascribed] to a most intelligent and designing agent. Whereas on the other hand, besides that the anomalies we speak of are incomparably fewer than those things which are regular and are produced in an orderly way; besides this, I say, the divine maker of the universe being a most free agent and having an intellect infinitely superior to ours, may in the production of seemingly irregular phenomena have ends unknown to us, which even the anomalies may be very fit to compass.

Thus when a man not versed in the mathematics looks upon a

curious geographical globe, though as soon as he perceives that the differing bignesses and particular confines of kingdoms and provinces and the apt situations, true distances and bearings of the cities and towns he knows by sight or fame be rightly set down, he cannot but conclude from these impresses of art or skill, that this was the work of a designing artificer. But though he also sees on the same globe several circles (as the tropics, the zodiac, the meridians, etc.), if he be a sober man he will not think that these were made by chance only, because he knows not the reasons or uses of them, or because some of the lines (as those curve[d] lines the seamen call rumbs) are not (like the other) circular, but do oddly and with a seeming irregularity intersect them. But [he] will rather think that the artist that had knowledge enough to represent the globe of the earth and waters in a body not two foot in diameter, had also skill enough to draw those lines with some design worthy of the same skill, though not obvious to those that are unacquainted with his art.

I did not incogitantly speak of irregularities, as if they might sometimes be but seeming ones. For I think it very possible that an artificer of so vast a comprehension and so piercing a sight as is the maker of the world might, in this great automaton of his, have so ordered things that divers of them may appear to us, and as it were break out abruptly and unexpectedly, and at great distances of time or place from one another, and on such accounts be thought irregular; which yet really have, both in his preordination and in the connection of their genuine causes, a reference that would, if we discerned it, keep us from imputing it either to chance or to nature's aberrations. To illustrate this a little, let us consider that if, when the Jesuits that first came into China presented a curious striking watch to the king, he that looked to it had wound up the alarm so as to strike a little after one; if (I say) this had been done, and that these Chinese that looked upon it as a living creature or some European animal, would think that when the index pointing at two of the clock likewise struck the same hour, and so three, four and onward, they would judge that these noises were regularly produced, because they (at equal intervals of time) heard them, and whensoever the index pointed at an hour, and never but then. But when the alarm came unexpectedly to make a loud, confused and more lasting noise, they could scarce avoid thinking that the animal was sick or exceedingly disordered. And yet the alarming noise did as properly flow from the structure of the little engine, and was as much designed by the manager of it, as those sounds of the clock that appeared manifestly regular.[m]

I foresee it may be said that unless we admit such a being as nature to contrive and manage things corporeal, and in a regular and methodical way direct them to their respective ends, there will appear no visible footsteps or proof of a divine wisdom in the corporeal world. And this argument, I confess, is so specious, that it was one of the things that made me the longest hesitate what I should think of the received notion of nature. But having further considered the matter, I saw it might be answered that the curious contrivance of the universe and many of its parts, and the orderly course of things corporeal with a manifest tendency to determinate ends, are matters of fact, and do not depend upon the supposition of such a being as they call nature, but (setting aside this or that hypothesis) may be known by inspection, if those that make the inspection be attentive and impartial: as, when a man sees a human body skilfully dissected by a dexterous anatomist, he cannot, if he be intelligent and unprejudiced, but acknowledge that there is a most curious and exquisite contrivance in that incomparable engine, and in the various parts of it that are admirably fitted for distinct and determinate functions or uses.

So that I do not at all – nor indeed can – suppress the manifest tokens of wisdom and design that are to be observed in the wonderful construction and orderly operations of the world and its parts, but I endeavour to refer these indications of wisdom to the true and proper cause. And whereas, in the hypothesis of the objectors, there may be three causes assigned of these specimens or footsteps of wisdom – namely, God, nature and chance – if, according to the doctrine by me proposed, nature be laid aside, the competition will remain only between God and chance. And sure he must be very dull, or very strongly prejudiced, that shall think it reasonable to attribute such admirable contrivances and such regular conducts as are observable in the corporeal world, rather to chance (which is a blind and senseless cause, or indeed no proper cause at all, but a kind of *ens rationis*[n]) than

---

[m] In the first edition, the rest of this section was printed out of order, at the end of section V rather than here.

[n] Scholastics distinguished between things which existed independently of our thinking about them and other entities which exist only in thought or as a result of our reasoning. An entity of the latter type is called an *en rationis*. Here Boyle may be implying that such entities are unreal or chimerical.

to a most intelligent being, from which the curiousest productions may with congruity be expected. Whereas, if such a celebrated thing as nature is commonly thought be admitted, it will not be near so easy to prove the wisdom (and consequently the existence) of God by his works, since they may have another cause – namely, that most watchful and provident being which men call nature. And this will be especially difficult in the Peripatetic hypothesis of the eternity (not of matter only, for in that the atomists and others agreed with them, but) of the world. For according to this account of the universe, there appears no necessity that God should have anything to do with it, since he did not make this automaton, but it was always self-existent, not only as to matter but to form too; and as for the government or administration of the bodies it consists of, that is the proper business of nature. And if it be objected, that this being is by its assertors acknowledged to be subordinate to God, I shall answer that, as upon the reasons and authorities I elsewhere deliver[11] it may justly be questioned whether many philosophers, and perhaps some sects of them who are adorers of nature, confessed her to be but the substitute of a superior and divine being? So this distinction and subordination is not so easy to be proved against those that side with those other ancient philosophers who either acknowledged no such thing or expressly denied it. Besides that, this objection supposes the existence and superiority of a deity, which therefore needs to be proved by other ways; whereas in the hypothesis I propose, the same phenomena that discover admirable wisdom and manifest designs in the corporeal world do themselves afford a solid argument, both of the existence and of some of the grand attributes of God, with which the rest that properly belong to him have a necessary connection.

[11] See the fourth section.

SECTION VII

I proceed now to the sixth [i.e. seventh] and difficultest part of my task, which is to show that the most general and current *effata* and axioms concerning nature that are wont to be employed in the writings of philosophers may have a fair account given of them, agreeably to the doctrine I have hitherto proposed, though these axioms do some of them suppose, and others seem strongly to support, the received notion of nature. To clear the way for the ensuing explications, I must desire you to recall to mind the two cautions I have formerly offered you (in the fifth section), wherewith I would have the common doctrine about the ends or designs of nature to be understood or limited.[a] And therefore I shall not here repeat what I there said, but only add in two words: that if those and some few other such things had been observed and duly considered, they might perhaps have prevented much of the obscurity and some of the errors that relate to the notion of nature.

I hope you have not forgot that the design of this paper was to examine the vulgar notion of nature, not to establish a new one of my own. And indeed the ambiguity of the word is so great (as has in the second section been made appear), and it is even by learned men frequently employed to signify such different things that, without enumerating and distinguishing its various acceptions, it were very unsafe to venture a giving a definition of it, and perhaps it were very impossible to give any that would not be liable to censure. I shall not therefore here presume to define a thing of which I have not found a stated and settled notion so far agreed on amongst men, but that I was obliged out of Aristotle and others to compile (in the fourth section) a collective representation of the vulgarly received idea or notion of nature, and afterwards to draw up as well as I could, instead of an accurate definition, tolerable descriptions of what on most occasions may be intelligibly meant by it. Wherefore, desiring and presuming that you will retain in your mind and, as occasion shall require, apply in the following part of this essay the things already delivered in the fourth section, I will not trouble you with the repetition of them.

But before I descend to treat of the particular *effata* or sentences that are received concerning nature's actings, it may not be improper nor unuseful to try if we can clear the way, by considering in what sense

[a]   These are given near the end of the fifth section, see pp. 73–5.

nature may or may not be said to act at all or to do this or that. For, for ought I can clearly discern, whatsoever is performed in the merely material world is really done by particular bodies acting according to the laws of motion, rest, etc. that are settled and maintained by God among things corporeal. In which hypothesis, nature seems rather a notional thing than a true physical and distinct or separate efficient [cause] such as would be, in case Aristotle's doctrine were true, one of those intelligences that he presumed to be the movers of the celestial orbs. But men do oftentimes express themselves so very ambiguously or intricately when they say that nature does this and that or that she acts thus and thus, that it is scarce (if at all) possible to translate their expressions into any forms of speech adequate to the original and yet intelligible. For which reason, though I have in the [fourth]^b section said something to the same purpose with what I am now to propose, yet the difficulty and weight of the subject makes me think it may be expedient, if not necessary, in this place somewhat more fully to declare what men do or should mean when they speak of nature's acting or of a thing's being naturally done or performed, by giving their words and phrases sometimes one interpretation and sometimes another.

1.   Sometimes when it is said that nature does this or that, it is less proper to say that it is done *by* nature than that it is done *according to* nature. So that nature is not here to be looked on as a distinct or separate agent, but as a rule or rather a system of rules according to which those agents and the bodies they work on are, by the great author of things, determined to act and suffer.

Thus, when water is raised in a sucking pump, it is said that nature makes the water ascend after the sucker to prevent a vacuum, though in reality this ascension is made not by such a separate agent as nature is fancied to be, but by the pressure of the atmosphere acting upon the water according to statical rules or the laws of^c the equilibrium of liquors settled by God among fluids, whether visible or pneumatical. So when the strict Peripatetics tell us that all the visible celestial orbs being by a motion that they call violent hurried about the earth every four and twenty hours from east to west, each of the planetary orbs has a natural motion that is quite contrary, tending from the west to the east; if they

---

^b   The section number was omitted by mistake. Presumably Boyle meant to refer to the fourth section, as he does in the previous paragraph.

^c   Here we have changed 'or' to 'of' for better sense.

will speak congruously to their master's doctrine, they must use the term 'natural' in the sense our observation gives it, since Aristotle will have the celestial orbs to be moved by external or separate agents – namely, spiritual intelligences.[d] Our observation may be also illustrated by other forms of speech that are in use: as when it is said, that the law takes care of infants and lunatics, that their indiscreet actions or omissions should not damnify their inheritances; and that the law hangs men for murder but only burns them in the hand for some lesser faults – of which phrases the meaning is that magistrates and other ministers of justice, acting according to the law of the land, do the things mentioned. And it tends yet more directly to our purpose to take notice that it is common to ascribe to art those things that are really performed by artificers, according to the prescriptions of the art; as when it is said that geometry (as the name imports) measures land, astrology foretells changes of weather and other future accidents, architecture makes buildings and chemistry prepares medicines.

2.   Sometimes when divers things such as the growth of trees, the maturations of fruits, etc., are said to be performed by the course of nature, the meaning ought to be that such things will be brought to pass by their proper and immediate causes, according to the wonted manner and series or order of their actings. Thus it is said that by the course of nature the summer days are longer than those of the winter; that when the moon is in opposition to the sun (that is, in the full moon), that part of her body which respects the earth is more enlightened than at the new moon or at either of the quadratures; and lastly, that when she enters more or less into the conical shadow of the earth, she suffers a total or a partial eclipse. And yet these and other illustrious phenomena may be clearly explicated without recourse to any such being as the Aristotelians' nature, barely by considering the situations and wonted motions of the sun or earth and the moon with reference to each other and to the terrestrial globe.

And here it may not be amiss to take notice that we may sometimes usefully distinguish between the laws of nature, more properly so called, and the custom of nature – or, if you please, between the fundamental and general constitutions among bodily things and the municipal laws (if I may so call them) that belong to this or that particular sort of bodies. As, to resume and somewhat vary our instance drawn from

---

[d]   Lat. lacks the phrase starting with 'since'.

water, when this falls to the ground, it may be said to do so by virtue of the custom of nature, it being almost constantly usual for that liquor to tend downwards and actually to fall down if it be not externally hindered. But when water ascends by suction in a pump or other instrument, that motion, being contrary to that which is wonted, is made in virtue of a more catholic law of nature, by which it is provided that a greater pressure (which in our case the water suffers from the weight of the incumbent air) should surmount a lesser (such as is here the gravity of the water that ascends in the pump or pipe).

The two foregoing observations may be further illustrated by considering in what sense men speak of things which they call preternatural, or else contrary to nature. For divers, if not most, of their expressions of this kind argue that nature is in them taken for the particular and subordinate or, as it were, the municipal laws established among bodies. Thus water, when it is intensely hot, is said to be in a preternatural state, because it is in one that it is not usual to it and, men think, does not regularly belong to it – though the fire or sun that thus agitates it and puts it into this state is confessed to be a natural agent, and is not thought to act otherwise than according to nature. Thus when a spring forcibly bent is conceived to be in a state contrary to its nature, as is argued from its incessant endeavour to remove the compressing body, this state, whether preternatural or contrary to nature, should be thought such but in reference to the springy body. For otherwise it is as agreeable to the grand laws that obtain among things corporeal that such a spring should remain bent by the degree of force that actually keeps it so, as that it should display itself in spite of a less or incompetent degree of force. And to omit the six non-natural things so much spoken of by physicians,[e] I must here take notice that, though a disease be generally reckoned as a preternatural thing or, as others carry the notion further, a state contrary to nature, yet that must be understood only with reference to what customarily happens to a human body – since excessively cold winds and immoderate rains and sultry air and other usual causes of diseases are as natural agents, and act as agreeably to the catholic laws of the universe, when they produce diseases, as when they condense the clouds into rain or snow, blow ships into their harbour, make rivers overflow, ripen corn and fruit, and

---

[e] A reference to the regimen based on human activities relating to air, diet, sleep, exercise, evacuation and passions of the mind.

do such other things, whether they be hurtful or beneficial to men. And upon a like account, when monsters are said to be preternatural things, the expression is to be understood with regard to that particular species of bodies from which the monster does enormously deviate, though the causes that produce that deviation act but according to the general laws whereby things corporeal are guided.

3. I doubt whether I should add as a third remark, or as somewhat that is referable to one or both of the two foregoing, that sometimes when it is said that nature performs this or that thing, we are not to conceive that this thing is an effect really produced by other than by proper physical causes or agents. But in such expressions, we are rather to look upon nature either as a relative thing, or as a term employed to denote a notional thing, with reference whereunto physical causes are considered as acting after some peculiar manner whereby we may distinguish their operations from those that are produced by other agents – or perhaps by the same, considered as acting in another way. This (I think) may be illustrated by some other received expressions or forms of speech. As when many of the ancient and some of the modern philosophers have said that things are brought fatally to pass, they did not mean that fate was a distinct and separate agent, but only that the physical causes performed the effect, as in their actings they had a necessary dependence upon one another or an inviolable connection that linked them together. And on the other side, when men say, as they too frequently do, that fortune or chance ἡ τυχὴ or τὸ αὐτόματον (for Aristotle[1] and his followers distinguish them, ascribing to the former what unexpectedly happens to deliberating or designing, and to the latter what happens to inanimate or undesigning, beings) has done this or that, considerate philosophers do not look upon fortune or chance as a true and distinct physical cause, but as a notional thing that denotes that the proper agents produced the effect without an intention to do so (as I have more fully declared in the fourth section).

One may (for ought I know) without impertinence refer to this our third observation, that many things are wont to be attributed to time; as when we say that time ripens some fruits that are too early gathered;

---

[1] *Differunt autem fortuna et casus, quia casus latius patet. Quod enim à fortuna est, casu est: hoc autem non omne est à fortuna.* [Fortune and chance are different, since chance is wider in scope. For what is due to fortune is also due to chance: but what is due to chance is not always due to fortune.] Aristotle, *Physics*, ii. 6. [The first edition refers to *Mirabiles auscultationes*: see p. 90 note 2.]

that it makes many things moulder and decay (*tempus edax rerum* [time the destroyer of things]); that it is the mother of truth; that it produces great alterations, both in the affairs of men and in their dispositions and their bodies – to omit many other vulgar expressions which represent time as the cause of several things, whereof really it is but an adjunct or a concomitant of the effects (however coincident with the successive parts of time and so someway related to it), being indeed produced by other agents that are their true and proper efficients.

Sometimes likewise, when it is said that nature does this or that, we ought not to suppose that the effect is produced by a distinct or separate being, but on such occasions the word 'nature' is to be conceived to signify a complex or convention of all the essential properties or necessary qualities that belong to a body of that species whereof the real agent is, or to more bodies respectively, if more must concur to the production of the effect. To this sense we are to expound many of those forms of speech that are wont to be employed when physicians or others speak of what nature does in reference to diseases or the cure of them. And to give a right sense to such expressions, I consider nature not as a principal and distinct agent, but a kind of compounded accident that is (as it were) made up of, or results from, the divers properties and qualities that belong to the true agents. And, that the name of a 'compounded accident' may not be startled at, I shall (to explain what I mean by it) observe that, as there are some qualities or accidents that (at least in comparison of others) may be called simple, as roundness, straightness, heat, gravity, etc.; so there are others that may be conceived as compounded, or made up of several qualities united in one subject – as in divers pigments, greenness is made up of blue and yellow, exquisitely mixed; beauty is made up of fit colours, taking features, just stature, fine shape, graceful motions and some other accidents of the human body and its parts. And of this sort of compounded accidents, I am apt to think there are far more than at the first mention of them one would imagine. And to this kind of beings, the expressions that naturists do on divers occasions employ incline me to think that what is called nature has a great affinity, at least in reference to those occasions. On which supposition one may conceive that, as when it is said that health makes a man eat well, digest well, sleep well, etc., considering [that] men do not look upon health as a distinct and separate cause of these effects, but as what we lately called a compounded accident – that is, a complex of all the real and genuine causes of good appetite, digestion, sleep, etc., in so much that health is not so properly the cause of these as their effect or result. So in divers things that nature is said to do, we need conceive no more than that the effects are produced by physical bodies and qualities or other proper causes, which, when we consider as conspiring or rather concurring to produce the same effect, by a compendious term we call nature.

By these and the like ways of interpretation, I thought fit to try whether I could give an intelligible and commodious sense to divers of the maxims or sentences and other forms of speech that are employed by those that on many occasions and in differing expressions say that nature does this or that, and acts thus and thus. But I confess that to clear all those ambiguous and unskilfully framed axioms and phrases, I found to be so intricate and difficult a task that, for want of time and perhaps too of patience, I grew weary before I had prosecuted it to the utmost. For which reason, though it is not improbable that some light may be given in this dark subject by what I have been now saying (as immature and unfinished as it is), especially if it be reflected on in conjunction with what has been formerly delivered (in the fourth section) about nature, general and particular; yet I shall at present make but very little use of the things that have been now said in expounding the axioms I am particularly to consider in this seventh section, hoping that I may, by the help of other mediums, dispatch my work without them. And to do it the more easily, I shall, without tying myself to the order wherein they are marshalled after the beginning of the fourth section, treat of them in the order wherein I think their explications may give most light to one another, or in that wherein the papers that belonged to them were retrieved.

[I.]   The first of the received axioms I shall consider is that which pronounces that *omnis natura est conservatrix sui* [all nature is its own conservator], where by the word 'nature' I suppose they understand a natural body, for otherwise I know not what they meant. Now this axiom easily admits of a twofold interpretation. For either it may signify no more than that no one body does tend to its own destruction, that is, to destroy itself; or else that in every body there is a principle called nature, upon whose score the body is vigilant and industrious to preserve its natural state and to defend itself from the violence and attempts of all other bodies that oppugn it or endeavour to destroy or harm it. In the former of these two senses, the axiom may be admitted without any prejudice to our doctrine. For since according to our hypothesis inanimate bodies can have neither appetites, nor hatreds, nor designs – which are all of them affections not of brute matter but of intelligent beings – I, that think inanimate bodies have no appetites at all, may easily grant that they have not any to destroy themselves. But according to the other sense of the proposed axiom, it will import that every body has within itself a principle whereby it does desire, and with all its power endeavour, to compass its own preservation; and both to do those things that tend thereunto, and oppose all endeavours that outward agents or internal distempers may use in order to the destruction of it. And as this is the most vulgar sense of this axiom, so it is chiefly in this sense that I am concerned to examine it.

I conceive, then, that the most wise creator of things did at first so frame the world and settle such laws of motion between the bodies that as parts compose it, that by the assistance of his general concourse the parts of the universe, especially those that are the greater and the more noble, are lodged in such places and furnished with such powers that, by the help of his general providence, they may have their beings continued and maintained as long and as far forth as the course he thought fit to establish amongst things corporeal requires. Upon this supposition, which is but a reasonable one, there will appear no necessity to have any recourse for the preservation of particular bodies to such an internal appetite and inbred knowledge in each of them, as our adversaries presume. Since, by virtue of the original frame of things and established laws of motion, bodies are necessarily determined to act on such occasions after the manner they would do if they had really an aim at self-preservation. As you see that if a blown bladder be compressed, and thereby the included air be forced out of its wonted dimensions and figure, it will incessantly endeavour to throw off and repel that which offers violence unto it, and first displace that part of the compressing body that it finds weakest, though in all this there be no appetite in the air (as I elsewhere show), no more than in the bladder, to that particular figure to maintain itself in which it seems so concerned.

Thus it is all one to a lump of dough, whether you make it into a round loaf, or a long roll, or a flat cake, or give it any other form. For whatever figure your hands or your instruments leave in it, that it will retain, without having any appetite to return to that which it last had.

So it is all one to a piece of wax, whether your seal imprints on it the figure of a wolf or that of a lamb. And for brevity's sake to pass by the instances that might be drawn from what happens to wood and marble and metals as they are differently shaped by the statuary's art and tools, I will only observe that the mariner's needle, before it is excited, may have no particular propensity to have respect to one part of heaven more than another. But when it has been duly touched upon a loadstone, the *flower-de-luce* will be determined to regard the north, and the opposite extreme the south. So that if the lily be drawn aside towards the east or towards the west, as soon as the force that detained it is removed, it will return to its former position and never rest until it regard the north. But in spite of this seeming affection of the lily to that point of the horizon, yet if the needle be duly touched upon the contrary pole of the same or another vigorous loadstone, the lily will presently forget its former inclination and regard the southern part of heaven – to which position it will, as it were, spontaneously return, having been forced aside towards the east or towards the west, if it be again left to its liberty. So that, though it formerly seemed so much to affect one point of heaven, yet it may in a trice be brought to have a strong propensity for the opposite, the lily having indeed no inclination for one point of heaven more than another, but resting in that position to which it was last determined by the prevalence of magnetical effluvia. And this example may serve to illustrate and confirm what we have been lately saying in general.

II.   Another received axiom concerning nature is that she never fails or misses of her end: *natura fine suo nunquam excidit*. This is a proposition whose ambiguity makes it uneasy for me to deliver my sense of it. But yet to say somewhat, if by 'nature' we here understand that being that the schoolmen style *natura naturans* [nature acting naturally], I grant or rather assert that nature never misses its end. For the omniscient and almighty author of things having once framed the world and established in it the laws of motion, which he constantly maintains, there can no irregularity or anomaly happen, especially among the greater mundane bodies, that he did not from the beginning foresee and think fit to permit, since they are but genuine consequences of that order of things that, at the beginning, he most wisely instituted – as I have formerly declared in instances of the eclipses of the sun and moon, to which I could add others, as the inundations of [the] Nile, so necessary to the health and plenty of Egypt. And though on some special occasions this instituted order, either seemingly or really, has been violated – as when the sun is said to have stood still in the days of Joshua, and the Red Sea to have divided itself to give free passage to the Israelites led by Moses[f] – yet these things having been rarely done for weighty ends and purposes by the peculiar intervention of the first cause either guiding or overruling the propensities and motions of secondary agents, it cannot be said that God is frustrated of his ends by these designed though seeming exorbitances by which he most wisely and effectually accomplishes them.

But if by 'nature' be meant such a subordinate principle as men are wont to understand by that name, I doubt the axiom is in many cases false. For though it be true, as I have often said, that the material world is so constituted that, for the most part, things are brought to pass by corporeal agents as regularly as if they designed the effects they produce, yet there are several cases wherein things happen quite otherwise. Thus it is confessed that when a woman is with child, the aim of nature is to produce a perfect or genuine human foetus; and yet we often see that nature, widely missing her mark, instead of that produces a monster. And of this we have such frequent instances that whole volumes have been published to recount and describe these gross and deformed aberrations of nature. We many times see (and have formerly noted) that in fevers and other acute diseases she makes critical attempts upon improper days, and in these unseasonable attempts does not only for the most part miss of her end, which is to cure the patient, but often brings him to a far worse condition than he was in before she used those miscarrying endeavours. To this may be referred the cheats men put upon nature, as when by grafting, the sap that nature raises with intention to feed the fruit of a white thorn (for instance) is by the gardener brought to nourish a fruit of quite another kind. So, when maltsters make barley to sprout, that germination whereby nature intended to produce stalks and ears is perverted to a far differing purpose, and she deluded. And now, to annex some arguments *ad hominem*, we are told that nature makes every agent aim at assimilating the patient to itself, and that upon this account, the fire aims at converting wood and the other bodies it works on into fire. But if this be so, nature must often miss of her end in chemical furnaces, where

f   See Joshua 10:12–14 and Exodus 14:21–2.

the flame does never turn the bricks that it makes red-hot into fire; nor the crucibles, nor the cupels,[g] nor yet the gold and silver that it thoroughly pervades and brings to be of a colour the same (or very near the same) with its own, and keeps in a very intense degree of heat and in actual fusion. And even when fire acts upon wood, there is but one part of it turned into fire, since, to say nothing of the soot and concreted smoke, the ashes remain fixed and incombustible. And so, to add another instance *ad hominem*, when we are told that nature makes water ascend in sucking pumps, *ob fugam vacui* [to avoid a vacuum], she must needs (as I formerly noted to another purpose) miss of her aim when the pump exceeds five and thirty, or forty, foot in height.[h] For then, though you pump never so much and withdraw the air from the upper part of the engine, the water will not ascend to the top, and consequently will leave a cavity, for whose replenishing she was supposed to have raised that liquor two or three and thirty foot.

III. Another of the celebrated axioms concerning nature is that she always acts by the shortest or most compendious ways: *natura semper agit per vias brevissimas*. But this rule, as well as divers others, does (I think) require to be somewhat explained and limited before it be admitted. For it is true that, as I have frequently occasion to inculcate, the omniscient author of the universe has so framed it that most of the parts of it act as regularly in order to the ends of it, as if they did it with design. But since inanimate bodies at least have no knowledge, it cannot reasonably be supposed that they moderate and vary their own actions according to the exigency of particular circumstances wherewith they must of necessity be unacquainted. And therefore it were strange if there were not divers occurrences wherein they are determined to act by other than the shortest ways that lead to particular ends, if those other ways be more congruous to the general laws or customs established among things corporeal.

This I prove by instances taken from gravity itself, which is perhaps that quality which of all others is most probably referred to an inbred power and propension. For it is true that if a stone or another heavy body be let fall into the free air, it will take its course directly towards the centre of the earth; and if it meet with an inclining plane, which

---

[g] Small vessels made of bone-ash used in assaying gold or silver with lead.

[h] Water will rise in a vacuum, under normal atmospheric pressure, to a height of about 34 feet (10 metres).

puts it out of its way, it will not for all that lose its tendency towards the centre, but run along that plane by which means its tendency downwards is prosecuted – though not, as before, in a perpendicular line, yet in the shortest way it is permitted to take. These obvious phenomena, I confess, agree very well with the vulgar axiom and possibly were the chief things that induced men to frame it. But now let us suppose that a small bullet of marble or steel, after having for a pretty space fallen through the air, lights upon a pavement of marble or some such hard stone that lies (as floors are wont to do) horizontal. In this case, experience shows (as was formerly noted on another occasion)[i] that the falling stone will rebound to a considerable height (in proportion to that it fell from) and, falling down again, rebound the second time, though not so high as before; and, in short, rebound several times before, by settling upon the floor, it approaches as near as is permitted it to the centre of heavy bodies. Whereas if nature did in all cases act by the most compendious ways, this bullet ought not to rebound at all, but, as soon as it found, by the hardness of the floor, it could descend no lower, it ought to have rested there, as in the nearest place it could obtain to the centre of the earth – whence every rebound must necessarily remove it to a greater distance. And so likewise, when a pendulum or bullet fastened to the end of a string is so held that the string is (*praeter propter* [more or less]) parallel[j] to the horizon, if it be thence let fall, it will not stop at the perpendicular line or line of direction which is supposed to reach from the nail or other prop through the centre of the bullet to the centre of the earth, but will pass beyond it and vibrate or swing to and fro, until it have passed again and again the line of direction for a great while, before the bullet come to settle in it – though, whenever it removes out of it towards either hand, it must really ascend or move upwards, and so go further off from the centre of the earth, to which, it is pretended, its innate propensity determines it to approach as much and as soon as is possible.[k] But this instance having been formerly touched upon,[l] I shall now observe to the same purpose that, having taken a good sea-compass [and the experiment succeeded with a naked, yet nicely poised, needle] and suffered the magnetic needle to rest north

---

i   See pp. 67–8. Lat. lacks this parenthetical phrase.
j   Here the first edition has 'perpendicular', but this is clearly erroneous.
k   Lat. lacks the next two sentences.
l   See p. 68. The square brackets in this sentence are Boyle's.

and south, if I held the proper pole of a good loadstone at a convenient distance on the right or left hand of the lily, this would be drawn aside from the north point towards the east or west as I pleased. And then the loadstone being removed quite away, the lily of the needle would indeed return northward, but would not stop in the magnetic meridian, but pass on divers degrees beyond it and would thence return without stopping at the meridian line; and so would, by its vibrations, describe many arches still shorter and shorter, until at length it came to settle on it and recover that position which – if nature always acted by the most compendious ways – it should have rested at the first time that by the removal of the loadstone it had regained it. But the truth is that at least inanimate bodies, acting without knowledge or design of their own, cannot stop or moderate their own action, but must necessarily move as they are determined by the catholic laws of motion – according to which, in one case, the impetus that the bullet acquires by falling is more powerful to carry it on beyond the line of direction than the action of the causes of gravity is to stop it as soon as it comes to the nearest place they can give it to the centre of the earth. And something like this happens in levity as well as gravity: for if you take an oblong and conveniently shaped piece of light wood (as fir or deal) and, having thrust or sunk it to the bottom of a somewhat deep stagnant water, give it liberty to ascend, it will not only regain the surface of the water – where, by the laws of gravity, it ought to rest (and did rest before it was forced down) – but it will pass far beyond that surface, and in part (as it were) shoot itself up into the incumbent air, and then fall down again and rise a second time (and perhaps much oftener) and fall again, before it come to settle in its due place, in which it is in an equilibrium with the water that endeavours to press it upwards.

IV.   Another of the sentences that are generally received concerning nature is that she always does what is best to be done: *natura semper quod optimum est facit.*[2] But of this it will not be safe for me to deliver my opinion until I have endeavoured to remove the ambiguity of the words, for they easily admit of two different senses, since they may signify that nature in the whole universe does always that which is best for the conservation of it in its present state, or that, in reference to each

---

[2]   *Natura semper id facit quod est optimum eorum quæ fieri possunt.* [Nature always does that which is the best of what can be done.] Aristotle, *De coelo,* ii. 4. See also Aristotle, *De generatione,* ii. 10 and 22.

body in particular, nature does still what is best – that is, what most conduces to the preservation and welfare of that body. If the first of these senses be pitched upon, the axiom will be less liable to exception. But then, I fear, it will be difficult to be positively made out by such instances as will prove that nature acts otherwise than necessarily according to laws mechanical; and therefore, until I meet with such proofs, I shall proceed to the other sense that may be given our axiom which, though it be the most usual, yet I confess I cannot admit without it be both explained and limited. I readily grant that the all-wise author of things corporeal has so framed the world that most things happen in it as if the particular bodies that compose it were watchful both for their own welfare and that of the universe. But I think withall that particular bodies, at least those that are inanimate, acting without either knowledge or design, their actions do not tend to what is best of them in their private capacities any further than will comport with the general laws of motion and the important customs established among things corporeal; so that to conform to these, divers things are done that are neither the best, nor so much as good, in reference to the welfare of particular bodies.

These sentiments I am induced to take up, not only by the more speculative considerations that have been formerly discoursed of and therefore shall not here be repeated, but by daily observations and obvious experience. We see oftentimes that fruit trees, especially when they grow old, will for one season be so overcharged with fruit that soon after they decay and die. And even while they flourish, the excessive weight of the too numerous fruits does not seldom break off the branches they grow upon, and thereby both hinders the maturity of the fruit and hastens the death of the tree: whereas this fatal profuseness would have been prevented if a wise nature, harboured in the plant, did (as is presumed) solicitously intend its welfare.

We see also, in divers diseases and in the unseasonable and hurtful crises of fevers, how far what men call nature oftentimes is from doing that which is best for the sick man's preservation. And indeed (as has been formerly noted on another occasion[m]) in many diseases (as bleedings, convulsions, choleras, etc.) a great part of the physician's work is to appease the fury and to correct the errors of nature – which being, as it were, transported with a blind and impetuous passion,

[m] See pp. 92–8.

unseasonably produces those dangerous disorders in the body that, if she were wise and watchful of its welfare, she would have been as careful to prevent as the physicians to remedy them.

Add to all this, that if nature be so provident and watchful for the good of men and other animals and of that part of the world wherein they live, how comes it to pass that from time to time she destroys such multitudes of men and beasts by earthquakes, pestilences, famine and other anomalies? And how comes it so often to pass in teeming women that, perhaps by a fright or a longing desire or the bare sight of any outward object, nature suffers herself to be so disordered and is brought to forget her plastic skill so much as, instead of well-formed infants, to produce hideous monsters, and those oftentimes so misshapen and ill contrived that not only themselves are unfit to live one day or perhaps one hour, but cannot come into the world without killing the mother that bare them. These and such other anomalies, though (as I have elsewhere shown)[n] they be not repugnant to the catholic laws of the universe and may be accounted for in the doctrine of God's general providence, yet they would seem to be aberrations, incongruous enough to the idea the schools give of nature, as of a being that, according to the axiom hitherto considered, does always that which is best. But it is time that we pass from that to the examen of another.

Though I have had occasion to treat of vacuum in the fifth section, yet I must also say something about it in this, because I there considered it but as it is employed by the Peripatetics and others to show the necessity of the principles they call nature.[o] But now I am to treat of it, not so much as an argument to be confuted, as on the force of its belonging to a (very plausible) axiom to be considered, although I fear that by reason of the identity of the subject (though considered in the fifth section and here, to differing purposes), I shall scarce avoid saying something or other co-incident with what has been said already.

V. The word 'vacuum' being ambiguous and used in differing senses, I think it requisite, before I declare my opinion about the generally received axiom of the schools that *natura vacuum horret* [nature abhors a vacuum], (or, as some express it, *abhorret a vacuo*), to premise the chief acceptions in which I have observed the term

---

[n] Boyle discusses the problem of explaining how God's attributes are compatible with evil on pp. 68–78.

[o] This paragraph was written in the 1680s; see MS 190, fol. 6v.

'vacuum' to be made use of. For it has sometimes a vulgar and sometimes a philosophical or strict signification. In common speech, 'to be empty' usually denotes not to be devoid of all body whatsoever, but of that body that men suppose should be in the thing spoken of, or of that which it was framed or designed to contain; as when men say that a purse is empty if there be no money in it; or a bladder, when the air is squeezed out; or a barrel, when either it has not been yet filled with liquor or has had the wine or other drink drawn out of it. The word 'vacuum' is also taken in another sense by philosophers that speak strictly, when they mean by it a space within the world (for I here meddle not with the imaginary spaces of the schoolmen, beyond the bounds of the universe[p]) wherein there is not contained any body whatsoever. This distinction being premised, I shall inform you that taking the word 'vacuum' in the strict sense, though many (and, among them, some of my best friends) pressed me to a declaration of my sense about the famous controversy *an detur vacuum* [whether a vacuum is obtained], because they were pleased to suppose I had made more trials than others had done about it, yet I have refused to declare myself either *pro* or *contra* [for or against] in that dispute, since the decision of the question seems to depend upon the stating of the true notion of a body, whose essence the Cartesians affirm and most other philosophers deny, to consist only in extension, according to the three dimensions length, breadth, and depth or thickness. For, if Monsieur Descartes' notion be admitted, it will be irrational to admit a vacuum, since any space that is pretended to be empty must be acknowledged to have the three dimensions, and consequently all that is necessary to essentiate a body.[q] And all the experiments that can be made with quicksilver or the *machina Boyliana* (as they call it),[r] or other instruments contrived for the like uses, will be eluded by the Cartesians, who will say that the space deserted by the mercury or the air is not empty, since it has length, breadth and depth, but is filled by their *materia subtilis* [subtle matter] that is fine enough to get freely in and out of the pores of the glasses,

---

[p] Boyle refers to the distinction between voids (or empty spaces) within the world (beneath the stellar sphere in traditional cosmology) and those outside the world (beyond the stellar sphere).

[q] According to Descartes, matter and extension (taking up space in three dimensions) were identical, indistinguishable concepts; hence there could be no 'spaces' without the presence of 'matter'.

[r] Boyle's air pump.

as the effluvia of the loadstone can do.[s] But though for these and other reasons I still forbear (as I lately said I have formerly done) to declare either way in the controversy about vacuum, yet I shall not stick to acknowledge that I do not acquiesce in the axiom of the schools that nature abhors a vacuum.

1.   For, first I consider that the chief, if not the only, reason that moves the generality of philosophers to believe that nature abhors a vacuum is that in some cases, as the ascension of water in sucking pumps, etc., they observe that there is an unusual endeavour, and perhaps a forcible motion in water and other bodies, to oppose a vacuum. But I – that see nothing to be manifest here, save that some bodies not devoid of weight have a motion upwards, or otherwise differing from their usual motions (as in determination, swiftness, etc.) – am not apt, without absolute necessity, to ascribe to inanimate and senseless bodies such as water, air, etc., the appetites and hatreds that belong to rational, or at least to sensitive, beings; and therefore think it a sufficient reason to decline employing such improper causes, if without them, the motions wont to be ascribed to them can be accounted for.

2.   If the Cartesian notion of the essence of a body be admitted by us, as it is by many modern philosophers and mathematicians, it can scarce be denied but that nature does not produce these oftentimes great, and oftener irregular, efforts to hinder a vacuum – since it being impossible there should be any, it were a fond thing to suppose that nature, who is represented to us as a most wise agent, should bestir herself and do extravagant feats to prevent an impossible mischief.

3.   If the atomical hypothesis be admitted, it must be granted not only that nature does not abhor a vacuum, but that a great part of the things she does require[s] it, since they are brought to pass by local motion. And yet there are very many cases wherein, according to these philosophers, the necessary motions of bodies cannot be performed unless the corpuscles that lie in their way have little empty spaces to retire, or be impelled, into when the body that pushes them endeavours

---

[s]   Descartes imagined that spaces apparently empty, such as the region above the meniscus in a mercury barometer, were in reality filled with 'subtle matter', extremely fine particles that could easily pass through the pores of tangible bodies. This 'subtle matter' also filled the heavens, and accounted for phenomena such as magnetism and gravitation. See his *Principles of Philosophy* (Amsterdam, 1644), part 2, art. 16–21 and part 3, art. 49–52.

to displace them. So that the *effatum* that nature abhors a vacuum agrees with neither of the two great sects of the modern philosophers.

But, without insisting on the authority of either of them, I consider that, for ought appears by the phenomena employed to demonstrate nature's abhorrence of a vacuum, it may be rational enough to think either that nature does not abhor a vacuum even when she seems solicitous to hinder it, or that she has but a very moderate hatred of it, in that sense wherein the vulgar philosophers take the word 'vacuum'.

For if we consider that, in almost all visible bodies here below and even in the atmospherical air itself, there is more or less of gravity or tendency towards the centre of our terraqueous globe, we may perceive that there is no need that nature should disquiet herself and act irregularly to hinder a vacuum, since without her abhorrence of it, it may be prevented or replenished by her affecting to place all heavy bodies as near the centre of the earth as [bodies] heavier than they will permit. And even without any design of hers, not to say without her existence, a vacuity will be as much opposed as we really find it to be, by the gravity of most (if not of all) bodies here below and the confluxibility of liquors and other fluids. For by virtue of their gravity and the minuteness of their parts, they will be determined to insinuate themselves into and fill all the spaces that they do not find already possessed by other bodies, either more ponderous in *specie* [in specific gravity] than themselves or, by reason of their firmness of structure, capable of resisting or hindering their descent. Agreeably to which notion we may observe that, where there is no danger of a vacuum, bodies may move as they do when they are said to endeavour its prevention – as, if you would thrust your fist deep into a pail full of sand and afterwards draw it out again, there will need nothing but the gravity of the sand to make it fill up the greatest part of the space deserted by your fist. And if the pail be replenished, instead of sand, with an aggregate of corpuscles more minute and glib than the grains of sand – as, for instance, with quicksilver or with water – then the space deserted by your hand will be, at least as to sense, completely filled up by the corpuscles of the liquor, which by their gravity, minuteness, and the fluidity of the body they compose, are determined to replenish the space deserted by the hand that was plunged into either of those liquors. And I elsewhere show[t] that if you take a pipe of glass whose cavity is too

[t]   Evidently in an unpublished work on hydrostatics.

narrow to let water and quicksilver pass by one another in it; if, I say, you take such a pipe and, having (by the help of suction) lodged a small cylinder of mercury of about half an inch long in the lower part of it, you carefully stop the upper orifice with the pulp of your finger, the quicksilver will remain suspended in the pipe. And if then you thrust the quicksilver directly downwards into a somewhat deep glass or other vessel full of water till the quicksilver be depressed about a foot or more beneath the surface of the water, if then you take off your finger from the orifice of the pipe which it stopped before, you shall immediately see the quicksilver ascend swiftly five or six inches and remain suspended at this new station. Which experiment seems manifestly to prove what I did long ago devise and do now allege it for, since here we have a sudden ascent of so heavy a body as is quicksilver and a suspension of it in the glass, not produced to prevent or fill a vacuum (for the pipe was open at both ends), the phenomena being but genuine consequences of the laws of the equilibrium of liquors, as I elsewhere clearly and particularly declare.

When I consider how great a power the school philosophers ascribe to nature, I am the less inclined to think that her abhorrence of a vacuum is so great as they believed. For I have shown in the fifth section that her aversion from it and her watchfulness against it are not so great, but that, in the sense of the Peripatetics, she can quietly enough admit it in some cases where, with a very small endeavour, she might prevent or replenish it, as I have particularly manifested in the forecited section. I just now mentioned a vacuum in the sense of the Peripatetics, because when the Torricellian experiment is made[u] – though it cannot perhaps be cogently proved either against the Cartesians or some other plenists that in the upper part of the tube, deserted by the quicksilver, there is a vacuum in the strict philosophical sense of the word – yet, as the Peripatetics declare their sense by divers of their reasonings against a vacuum mentioned in that section, it will to a heedful peruser appear very hard for them to show that there is not one in that tube. And, as by the schoolmen's way of arguing nature's hatred of a vacuum from the suspension of water and other liquors in tubes and conical watering posts, it appears that they thought that any

[u] The experiment first performed by Evangelista Torricelli (1608–47) in 1644 by which a tube sealed at one end and filled with mercury was inverted in a dish of the same substance, creating a space at the top of the tube.

space here below deserted by a visible body, not succeeded by another visible body or at least by common air, may be reputed empty. So, by the space deserted by the quicksilver at the top of the pipe of a baroscope thirty-one inches long, one may be invited to doubt whether a vacuum ought to be thought so formidable a thing to nature, as they imagine she does and ought to think it. For what mischief do we see ensue to the universe upon the producing or continuance of such a vacuum, though the deserted space were many time[s] greater than an inch and continued many years, as has divers times happened in the taller sort of mercurial baroscopes? And those Peripatetics that tell us that, if there were a vacuum, the influences of the celestial bodies that are absolutely necessary to the preservation of sublunary ones would be intercepted, since motion cannot be made *in vacuo* [in a vacuum], would do well to prove, not suppose, such a necessity; and also to consider that, in our case, the top of the quicksilver to which the vacuum reaches does usually appear protuberant, which shows that the beams of light (which they think of great affinity to influences, if not the vehicle) are able to traverse that vacuum, being in spite of it reflected from the mercury to the beholder's eye. And in such a vacuum, as to common air, I have tried that a loadstone will emit his effluvia and move iron or steel placed in it.

In short, it is not evident that here below nature so much strains herself to hinder or fill up a vacuum as to manifest an abhorrence of it. And without much peculiar solicitude, a vacuum, at least a philosophical one, is as much provided against as the welfare of the universe requires, by gravity and confluxibility of the liquors and other bodies that are placed here below. And as for those that tell us that nature abhors and prevents a vacuum as well in the upper part of the world as the lower, I think we need not trouble ourselves to answer the allegation till they have proved it – which I think will be very hard for them to do, not to mention that a Cartesian may tell them that it were as needless for nature to oppose a vacuum in heaven as in earth, since the production of it is everywhere alike impossible.[v]

VI. I come now to the celebrated saying that *natura est morborum medicatrix* [nature is the curer of diseases], taken from Hippocrates, who expresses it in the plural,[3] νουσων φυσεις ἰητροι [natures are the

---

[3]   Hippocrates, *Epidemics*, lib. 6. § 5. t[ext]. 1.

[v]   Drafts written *c.* 1680 survive for most of the rest of this section, except the final three paragraphs; see BP 18, fols. 105–9; MS 198, fols. 3–8; and MS 199, fols. 147–50.

curers of diseases]. And because this axiom is generally received among physicians and philosophers, and seems to be one of the principal things that has made them introduce such a being as they call nature, I think it may be time well employed to consider somewhat attentively in what sense and how far this famous sentence may or should not be admitted.

First then, I conceive it may be taken in a negative sense, so as to import that diseases cannot be cured in such persons in whom' the aggregate of the vital powers or faculties of the body is so far weakened or depraved as to be utterly unable to perform the functions necessary to life, or at least to actuate and assist the remedies employed by the physician to preserve or recover the patient. This I take to be the meaning of such usual phrases as, that 'physic comes too late' and that 'nature is quite spent'. And in this sense, I readily acknowledge the axiom to be true. For where the engine has some necessary parts, whether fluid or solid, so far depraved or weakened as to render it altogether unable to co-operate with the medicine, it cannot be rationally expected that the administration of that medicine should be effectual. But in this (I presume) there is no difficulty worthy to detain us.

I proceed therefore to the positive sense whereof our axiom is capable, and wherein it is the most usually employed. For men are wont to believe that there resides in the body of a sick person a certain provident or watchful being that still industriously employs itself by its own endeavours, as well as by any occasional assistance that may be afforded it by the physician, to rectify whatever is amiss and restore the distempered body to its pristine state of health. What I think of this doctrine I shall leave you to gather from the following discourse.

I conceive then in the first place, that the wise and beneficent maker of the world and of man, intending that men should, for the most part, live a considerable number of years in a condition to act their parts on the mundane stage, he was pleased to frame those living automatons, human bodies, that – with the ordinary succours of reason, making use of their exquisite structure fitted for durableness, and of the friendly, though undesigned, assistance of the various bodies among which they are placed – they may in many cases recover a state of health, if they chance to be put out of it by lesser accidents than those that God, in compliance with the great ends of his general providence, did not think fit to secure them from or enable them to surmount. Many things therefore that are commonly ascribed to nature, I think, may be better ascribed to the mechanisms of the macrocosm and microcosm – I mean, of the universe and the human body.

And to illustrate a little my meaning by a gross example or two, I desire you will consider with me a sea-compass wherein the excited magnetic needle and the box that holds it are duly poised by means of a competent number of opposite pivots. For though, if you give this instrument a somewhat rude shake, you will make the box totter and incline this way and that way, and at the same time drive the points of the magnetic needle many degrees to the east or to the west, yet the construction of the instrument and the magnetism of one main part of it are such that, if the force that first put it into a disorderly motion cease from acting on it, the box will after some reciprocations return to its horizontal situation; and the needle that was forced to deviate will, after a few irregular motions to this and to that side of the magnetical meridian, settle itself again in the position, wherein the *flower-de-luce* steadfastly regards the north. And yet this recovery to its former state is effected in a factitious body by the bare mechanism of the instrument itself and of the earth and other bodies within whose sphere of activity it is placed.

But, because many have not seen a mariner's compass, I will add a less apposite but more obvious and familiar example. For if when an empty balance is duly counterpoised, you shall by your breath or hand depress one of the scales and thereby, for the time, destroy the equilibrium, yet when the force is once removed, the depressed balance will presently ascend and the opposite will descend. And after a few motions up and down, they will both of them of their own accord settle again in an exact equilibrium, without the help of any such provident internal principle as nature – the absence of whose agency may be confirmed by this, that the depressed scale does not at first stop at the horizontal line beneath which it was first depressed (as it ought to do if it were raised by an intelligent being), but rises far above it.[w]

If it be here objected that these examples are drawn from factitious, not from merely physical, bodies, I shall return this brief answer and desire that it be applied not only to the two freshly mentioned examples, but to all of the like kind that may be met with in this whole treatise. I

---

[w] Lat. lacks the following paragraph and part of the next sentence, up to 'thus in a human body' (p. 127 l. 14).

say then, in short, that divers of the instances we are speaking of are intended but for illustrations, and that others may be useful instances if they should be no more than analogous ones, since examples drawn from artificial bodies and things may have both the advantage of being more clearly conceived by ordinary understandings, and that of being less obnoxious to be questioned in that particular in which the comparison or correspondence consists. And I the less scruple to employ such examples, because Aristotle himself and some of his more learned followers make use of divers comparisons drawn from the figures and other accidents of artificial things to give an account of physical subjects and even of the generation, corruption and forms of natural bodies.

This advertisement premised, I pursue this discourse it interrupted by adding: thus in a human body, the causes that disorder it are oftentimes but transient, whereas the structure of the body itself and the causes that conduce to the preservation of that structure are more stable and durable, and on that account may enable the engine to outlast many things that are hostile to it. This may be somewhat illustrated by considering that sleep, though it be not properly a disease, easily becomes one when it frequently transgresses its due bounds; and even while it keeps within them, it does for the time it lasts hinder the exercise of many functions of the body more than several diseases do; and yet, according to the common course of things, the matter that locked up the senses being spent, the man of himself recovers that sensible and active state on whose score he is said to be awake. But to come somewhat closer to the point, we see that many persons who get a preternatural thirst with overmuch drinking, get rid of it again in a few days by forbearing such excesses. And many, that by too plentiful meals are brought to a want of appetite, recover (as it were), of course, by a spare diet in a few days – the renewed ferment, or menstruum of the stomach, being able in that time to concoct by little and little, or expel the indigested aliments or peccant humours that offended the stomach and caused the want of appetite.

And here I desire to have it taken notice of, as a thing that may be considerable to our present purpose, that I look not on a human body as on a watch or a hand mill – i.e. as a machine made up only of solid or at least consistent parts – but as an hydraulical, or rather hydraulo-pneumatical, engine, that consists not only of solid and stable parts, but of fluids and those in organical motion. And not only so, but I consider that these fluids, the liquors and spirits, are in a living man so constituted, that in certain circumstances the liquors are disposed to be put into a fermentation or commotion whereby either some depuration of themselves, or some discharge of hurtful matter by excretion, or both, are produced so as for the most part to conduce to the recovery or welfare of the body.

And that even consistent parts may be so framed and so connected with other parts as to act, as it were, *pro re nata* [for the benefit of the organism], varying their motions as differing circumstances make it convenient they should be varied, I purposely show in another paper.[x] To this I might altogether refer you, but in regard [that] the thing is a paradox and lays a foundation for another not inferior to itself, I shall here borrow thence one instance, not mentioned that I know of by others to this purpose, that may both declare my meaning and confirm the thing itself. I consider then that what is called the pupil or apple of the eye is not (as it is known) a substantial part of the organ, but only a round hole or window made in the uvea, at which the modified beams of light enter, to fall upon the crystalline humour and thence be refracted to the bottom of the eye, or seat of vision, to make there an impression that is usually a kind of picture (for it is not always a neat one) of the object. Now the wise and all-foreseeing author of things has so admirably contrived this instrument of sight that, as it happens to be employed in differing lights, so the bigness or area of the pupil varies. For when the light is vivid and would be too great if all the beams were let in that might enter at an aperture as large as the usual, the curtain is every way drawn towards the middle, and thereby the round window made narrower. And on the other side, when the light is but faint and the object but dimly illustrated, there being more light requisite to make a sufficient impression at the bottom of the eye, the curtain is every way drawn open to let in more light. And when the eye is well constituted, this is regularly done, according as the organ has need of more or less light. Of this, some late masters of optics have well treated, and I have spoken about it more fully in another place.[y] And the truth of the

[x] See *A Disquisition about the Final Causes of Natural Things* (London, 1688).
[y] See *A Disquisition about the Final Causes*, pp. 148–9 (*Works*, vol. 5, pp. 425f.); and 'A Sceptical Dialogue about the Positive and Privative Nature of Cold', in *Tracts Consisting of Observations about the Saltness of the Sea* (Oxford, 1674), p. 20 (*Works*, vol. 3, p. 741).

observation you may easily find, if you look upon the eyes of a boy or a girl (for in young persons the change is the most notable) when the eyes are turned from looking on dark objects towards bright or more illuminated ones. And I have found the variation yet more conspicuous in the eyes of a young cat, as I elsewhere particularly relate.[z] So that, referring you to the writings already pointed at, I shall only add in this place that these various motions in the eye are produced by mere mechanism, without the direction, or so much as knowledge or perception, of the rational soul. And upon the like account it is that other motions, in several parts belonging to the eye, are produced (as it were) spontaneously, as occasion requires. And so as to the fluid parts of the body, we find that, according to the institution of the author of things, when healthy women are of a fit age, there is a monthly fermentation or commotion made in the blood, which usually produces a kind of separation, and then an excretion, advantageous to the body.

And that you may the better make out what I meant by the disposition or tendency of the parts to return to their former constitution, I shall desire you to consider with me a thin and narrow plate of good steel or refined silver. For if one end of it be forcibly drawn aside, the changed texture of the parts becomes such – or the congruity and incongruity of the pores, in reference to the ambient ether that endeavours to permeate them, is made such – that as soon as the force that bent it is removed, the plate does (as it were) spontaneously return to its former position. And yet here is no internal watchful principle that is solicitous to make this restitution, for otherwise it is indifferent to the plate what figure it settle in. For if the springy body stand long bent, then – as if nature forgot her office or were unable to execute it – though the force that held the spring bent be removed, it will not endeavour to regain its former straightness. And I have tried in a silver plate that, if you only heat it red-hot and let it cool, if you put it into a crooked posture, it will retain it; but barely with two or three strokes of a hammer, which can only make an invisible change of texture, the plate will acquire a manifest and considerable springiness, which you may again deprive it of by sufficiently heating it in the fire without so much as melting it.

But, to return to the discourse formerly begun about distempers wont to be harmless by being transient, we may observe that the third

[z]   We have not located this reference.

or fourth day after women are brought to bed, there is commonly a kind of fever produced upon the plentiful resort of the milk to the breasts, for which cause this distemper is by many called the fever of milk. And this is wont in a short time to pass away of itself, as depending upon causes far less durable than the economy of the woman's body. And if it be objected that these are not diseases because they happen according to the instituted course of nature, I will not now dispute the validity of the consequence, though I could represent that the labour of teeming women and the breeding of teeth in children happen as much according to the institution of nature, and yet are usually very painful and oftentimes dangerous. But I will rather answer that, if the troublesome accidents I have alleged cannot serve to prove, they may at least to illustrate, what I aim at. And I shall proceed to take notice of a distemper that physicians generally reckon among diseases – I mean the flowing of blood at the haemorrhoidal veins. For though oftentimes this flux of blood is excessive and so becomes very dangerous, and therefore must be checked by the physician (which is no great argument that a being wise and watchful manages this evacuation), yet frequently, if not for the most part, the constitution of the body is such that the superfluous or vitiated blood goes off before it has been able to do any considerable mischief, or perhaps any at all, to the body. And so we see that many coughs and hoarsenesses and coryzas are said to be cured – that is, do cease to trouble men – though no medicine be used against them, the structure of the body being durable enough to outlast the peccant matters or the operation of those other causes that produce these distempers.

It is a known thing that most persons, the first time they go to sea, especially if the weather be anything stormy, are by the unwonted agitations which those of the ship produce in them (assisted perhaps by the sea-air and smells of the ship) cast into that disease that, from the cause of it, is called the seasickness, which is sometimes dangerous and always very troublesome, usually causing a loss of appetite and almost continual faintness, a pain in the head and almost constant nauseousness, accompanied with frequent and oftentimes violent vomitings – which symptoms make many complain that, for the time, they never felt so troublesome a sickness. And yet usually after not many days, this distemper by degrees is mastered by the powers of the body, tending still to persevere in their orderly and friendly course and suppressing the adventitious motions that oppose it, and the sick person recovers without other help. And so, though persons unaccustomed to the sea, whether they be sick or no, are by the inconvenient motions of the ship usually brought to a kind of habitual giddiness, which disposes them to reel and falter when they walk upon firm ground, yet when they come ashore, they are wont in no long time to be freed from this uneasy giddiness without the help of any medicine – the usual and regular motions of the parts of the body obliterating by degrees in a few days (I used to be free from it within some hours) that adventitious impression that caused the discomposure.

To the same purpose we may take notice of that which happens to many persons who, riding backwards in a coach, are not only much distempered in their heads, but are made very sick in their stomachs and forced to vomit as violently and frequently as if they had taken an emetic. And yet all this disorder is wont quickly to cease when the patient leaves the coach, without the continuance of whose motion (that continues a preposterous one in some parts of the patient), the distemper will quickly yield to the more ordinary and regular motions of the blood and other fluids of the body. So when in a coach or elsewhere, a man happens to be brought to faintness or cast into a swoon by the closeness of the place or the overcharging of the air with the fuliginous reeks of men's bodies, though the disease be formidable, yet if the patient be seasonably brought into the free air, the friendly operation of that external body, assisting the usual endeavours or tendency of the parts of the patient's body to maintain his life and health, is wont quickly to restore him to the state he was in before this sudden sickness invaded him.

Divers things that happen in some diseases may be grossly illustrated by supposing that into a vial of fair water some mud be put and then the vial be well shaken, for the water will be troubled and dirty and will lose its transparency upon a double account: that of the mud, whose opacous particles are confounded with it, and that of the newly generated bubbles that swim at the top of it. And yet to clarify this water and make it recover its former limpidness, there needs no particular care or design of nature, but according to the common course of things, after some time the bubbles will break and vanish at the top, and the earthy particles that compose the mud will by their gravity subside to the bottom and settle there, and so the water will become clear again. Thus also must, which is the lately expressed juice of grapes, will for a good while continue a troubled liquor; but though there be no substantial form to guide the motions of this factitious body, yet according to the course of things, a fermentation is excited and some corpuscles are driven away in the form of exhalations or vapours, others are thrown against the sides of the cask and hardened there into tartar, and others again subside to the bottom and settle there in the form of lees, and by this means leave the liquor clear and (as to sense) uniform. And why may not some depurations and proscriptions of heterogeneous parts be made in the blood as well as they are usually in must, without any peculiar and solicitous director of nature?

There is indeed one thing to which the sentence of 'nature's being the curer of diseases' may be very speciously applied, and that is the healing of cuts and wounds, which, if they be but in the flesh, may oftentimes be cured without plasters, salves or other medicines. But, not to mention haemorrhages and some other symptoms wherein the surgeon is fain to curb or remedy the exorbitancies of nature, this healing of the *solutio continui*[a] seems to be but an effect or consequent of that fabric of the body on which nutrition depends. For the alimental juice being by the circulation of the blood and chyle carried to all parts of the body to be nourished, if it meets anywhere either with preternatural concretions, or with a gap made by a cut or wound, its particles do there concrete into a kind of bastard flesh or some such other body, which that juice, in the place and other circumstances it is in, is fitted to constitute. Thus we see that not only wens and scrofulous tumours are nourished in the body, but misshapen moles do by nutriment grow in the womb, as well as embryos feed there. And to come closer to the present argument, we see that in wounds, proud flesh and perhaps fungi are as well produced and entertained by the aliment brought to the wounded part as the true and genuine flesh; so that either nature seems much mistaken, if she designs the production and maintenance of such superfluous and inconvenient bodies, or the surgeon is much to blame, who is industrious to destroy them, though oftentimes he cannot do it without using painful corrosives. But for ought appears, nature is not so shy and reserved in her bounty, but that

---

[a] According to Stephen Blancard [Blankaart], *A Physical Dictionary*, trans. J. G. (London, 1684): 'a Dissolution of the Unity and Continuity of the Parts: As in Wounds, Ulcers, Fragments, etc.'; i.e. breaking a bone or cutting the skin disrupts the unity of a part.

she sends nourishment to repair as well things that do not belong to the body as genuine parts of it, as to restore flesh to wounded parts as may appear by warts and corns that grow again after they are not skilfully cut.[b] And I remember I have seen a woman in whose forehead nature was careful to nourish a horn about an inch and more in length, which I fully examined while it was yet growing upon her head, to avoid being imposed upon.

But besides the diseases hitherto discoursed, there are many others, as well acute as chronical, wherein it is confessed that nature alone does not work the cure, so that as to these (which are more numerous than the former), I may well pretend that the aphorism that makes nature the curer of diseases is not true, otherwise than in a limited sense. But because I know it is pretended that even in these diseases nature is the principal agent by whose direction the physician acts in subserviency to her designs, and physicians themselves (whether out of modesty or inadvertence, I now enquire not) are wont to acknowledge that they are but nature's ministers, I think it necessary to consider briefly what sense is fit, according to our doctrine, to be given to these assertions, to make them receivable by us.

But to make way for what we are to say on this occasion, it may be fit to observe, that one great cause of the common mistakes about this matter is (as has been partly intimated already) that the body of a man is looked upon rather as a system of parts, whereof most are gross and consistent, and not a few hard and solid too, than as what indeed it is: a very compounded engine that, besides these consistent parts, does consist of the blood, chyle, gall and other liquors; also of more subtle fluids, as spirits and air; all which liquors and fluids are almost incessantly and variously moving, and thereby put divers of the solid parts, as the heart and lungs, the diaphragm, the hands, feet, etc., into frequent and differing motions. So that, as when the constitution or the motions, that in a sound body do regularly belong to the fluid parts, happens the former to be depraved or the latter to grow anomalous, the engine is immediately out of order, though the gross and solid parts[c] were not primarily affected; so, when by proper remedies (whether visible or not) the vitiated texture or crasis of the blood or other juices is

---

[b] Here the first edition has 'are skilfully cut', but we follow the original draft (BP 18, fol. 108r), which makes more sense.

[c] Here we follow the original draft (BP 18, fol. 108v); the first edition has 'gross solid parts'.

corrected, and the inordinate motions that they and the spirits are put into (or that they also put the consistent parts into) are calmed and rectified, the grosser and more solid parts of the body – and so the whole animal economy, if I may so call it – will be restored to a more convenient state. Thus we see, that in many hysterical women, by the fragrant effluvia of a Spanish glove or some other strong perfume, the spirits and *genus nervosum*[d] being affected, several disorderly symptoms are produced, and oftentimes the motion of the blood is so stopped or abated that any pulse at all is scarcely to be felt, nor respiration discerned; and the whole engine, unable to sustain itself, falls to the ground and lies moveless on it. And yet we have often, by barely holding to the patient's nostrils a vial full of very strong spirit or volatile salt or sal-armoniac or of harts-horn,[e] in less than a quarter of an hour, sometimes in a few minutes, restored women in that condition to their senses, speech and motion.

We are also here to consider what I have formerly inculcated: that the economy of the human body is so constituted by the divine author of it, that it is usually fitted to last many years, if the more general laws settled by the same author of the universe will permit it. And therefore it is not to be wondered at, that in many cases the automaton should be in a condition to concur, though not with knowledge and design, to its own preservation, when, though it had been put somewhat out of order, it is assisted by the physician's hands or medicines to recover a convenient state.

And if it be objected that the examples that have been in this past discourse frequently drawn from automata are not adequate and do not fully reach the difficulties we have been speaking of, I shall readily grant it, provided it be considered that I avowedly and deservedly suppose the bodies of living animals to be, originally, engines of God's own framing and consequently effects of an omniscient and almighty artificer. So that it is not rational to expect that, in the incomparably inferior productions of human skill, there should be found engines fit to be compared with these which, in their protoplasts, had God for their author.[f] Not to mention (what yet may be considerable in reference to the lastingness of

---

[d]   'Stuff of the nerves', apparently a reference to the nervous system.
[e]   Boyle here refers to various ammonia-based compounds: a salt of ammonia, probably the carbonate; ammonium chlorate; and ammonia water.
[f]   Lat. lacks the rest of this paragraph.

human life) that a man is not a mere mechanical thing, where nothing is performed for the preservation of the engine or its recovery to a good state but by its own parts or by other agents, acting according to mechanical laws without counsel or design; since, though the body of a man be indeed an engine, yet there is united to it an intelligent being (the rational soul or mind), which is capable – especially if instructed by the physicians' art to discern – in many cases, what may hurt it and what may conduce to the welfare of it, and is also able (by the power it has to govern the muscles and other instruments of voluntary motion) to do many of those things it judges most conducive to the safety and the welfare of the body it is joined with. So that a man is not like a watch or an empty boat, where there is nothing but what is purely mechanical, but like a manned boat, where, besides the machinal part (if I may so speak), there is an intelligent being that takes care of it, and both steers it or otherwise guides it and, when need requires, trims it; and, in a word, as occasion serves, does what he can to preserve it and keep it fit for the purposes it is designed for.

These things being premised, I think the physician (here supposed to be free from prejudices and mistakes) is to look upon his patient's body as an engine that is out of order, but yet is so constituted that, by his concurrence with the endeavours, or rather tendencies, of the parts of the automaton itself, it may be brought to a better state. If therefore he find that, in the present disposition of the body, there is a propensity or tendency to throw off the matter that offends it, and (which ought to be some way or other expelled) in a convenient way and at commodious places, he will then act so as to comply with and further that way of discharge rather than another. As, if there be a great appearance that a disease will quickly have a crisis by sweat, he will rather further it by covering the patient with warm clothes and giving sudorific medicines than, by endeavouring to carry off the peccant matter by purging or vomiting, unseasonably hinder a discharge that probably will be beneficial. And in this sense men may say, if they please, that the physicians are ministers or servants of nature; as seamen, when the ship goes before a good wind, will not shift their sails, nor alter the ship's motion, because they need not. But to show that it is (as it were) by accident that the physician does, in the forementioned case, obey nature (to speak in the language of the naturists I reason with), I need but represent that there are many other cases wherein the physician, if he be skilful, will be so far from taking nature for his mistress to direct him by her example what should be done, that a great part of his care and skill is employed to hinder her from doing what she seems to design, and to bring to pass other things very differing from, if not contrary to, what she endeavours.

Thus though nature in dropsies importunately crave[s] store of drink, the physician thinks himself obliged to deny it; as he does what they greedily desire, to his patients of the green sickness, or that distemper they call pica, though the absurd and hurtful things, as very unripe fruit, lime, coals and other incongruous things, be earnestly longed for. Thus also the surgeon does often hinder nature from closing up the lips of a wound, as she would unskilfully do, before it be well and securely healed at the bottom. So the physician does often, by purging or phlebotomy, carry off that matter that nature would more dangerously throw into the lungs and expel by frequent and violent coughs.

And so if a nerve or tendon be pricked, the surgeon is fain, with anodynes and other convenient medicines, to prevent or appease the unreasonable transports of nature when, being in a fury, by violent and threatening convulsions she not only much disorders, but endangers the patient. And so likewise when in those evacuations that are peculiar to women, nature affects in some individuals to make them by undue and inconvenient places, as the nipples, the mouth or the eyes, whereof we have divers instances among the observations collected by Schenckius or related by other good authors.[4] The physician is careful by bleeding the patient in the foot and by using other means to oblige nature to alter her purpose and make the intended evacuations by the proper uterine vessels. And though according to the institution of nature, as they speak, there ought to be a monthly discharge of these superfluities, and therefore, while this is moderately made, the physician does rather further than suppress it: yet if, as it often happens in other patients, nature overlashes in making those evacuations, to the great weakening or endangering the sick person, the physician is careful by contemperating medicines and other ways to correct nature's exorbitancy and check her profuseness of so necessary a liquor as the blood.

Other instances more considerable than some of these hitherto mentioned might be given to the same purpose, but I forbear to do it

---

[4] Schenk. Obser. l[iber]. IV. pag. m. 633. & seq. ['pages 633f. in my copy'; citing *Observationum medicarum rerum* (Frankfurt, 1609), by the German physician Johann Schenck (1530–98).]

because there being some, though perhaps very needless, controversies about them, I could not make out their fitness to be here alleged without more words than I am now willing to employ about unnecessary proofs, fearing it might be thought I have dwelt too long already upon the explication of one aphorism. I shall therefore only observe in short, that I look upon a good physician not so properly as a servant to nature, as one that is a counsellor and a friendly assistant, who in his patient's body furthers these motions and other things that he judges conducive to the welfare and recovery of it. But as to those that he perceives likely to be hurtful, either by increasing the disease or otherwise endangering the patient, he think[s] it is his part to oppose or hinder, though nature do manifestly enough seem to endeavour the exercising or carrying on those hurtful motions.

On this occasion, I shall take notice of the practice of the more prudent among physicians themselves who, being called to a patient subject to the flux of the haemorrhoids, if they find the evacuation to be moderate and likely either to benefit the patient on another account (as in some cases it is) or at least to end well, they do, as some of them speak, commit the whole business to nature – that is, to speak intelligibly, they suffer it to take its course, being encouraged to do so in some cases by the doctrine of Hippocrates[5] and in others by experience. But if the evacuation prove to be too lasting or too copious, they then are careful to hinder nature from proceeding in it and think themselves obliged to employ both inward and outward means to put a stop to an evacuation, which may bring on a dropsy or some other formidable disease. And if it be said that nature makes this profusion of so necessary a liquor as blood only because she is irritated by the acrimony of some humour mixed with it, I say that this answer – which, for substance, is the same that naturists may be compelled to fly to on many occasions – is in effect a confession that nature is no such wise being as they pretend, since she is so often provoked to act (as it were) in a fury and do those things in the body that would be very mischievous to it, if the physician, more calm and wise than she, did not hinder her.[g] So that, notwithstanding the reverence I pay the great Hippocrates, it is not without due caution and some limitations that I admit that notable

---

5 Hippocrat. Lib. vi. Aphorism. xi. [Citing Hippocrates, *Aphorismi.*]

g Lat. lacks the rest of this paragraph.

sentence of his, where he thus speaks: *invenit Natura ipsa sibi-ipsi aggressiones.* [Nature herself finds her own modes of approach.] And after three or four lines, *non edocta Natura et nullo magistro usa, ea quibus opus est facit.*[6] [Nature, untaught and lacking an instructor, does what is needed.] Which, I fear, makes many physicians less courageous and careful than they should, or perhaps would be, to employ their own skill on divers occasions that much require it.

I shall now add that, as in some cases the physician relieves his patient in a negative way by opposing nature in her unseasonable or disorderly attempts, so in other cases he may do it in a positive way[h] by employing medicines that either strengthen the parts, as well fluid as stable, or make sensible evacuations of matters necessary to be proscribed by them, or he may do it by using remedies that by their manifest qualities oppugn those of the morbific matter or causes; as when, by alkalis or absorbing medicaments, he mortifies preternatural acids or disables them to do mischief. And perhaps one may venture to say that in some cases the physician may in a positive way contribute more to the cure even of an inward disease than nature herself seems able to do. For if there be any such medicine preparable by art, as Helmont affirms may be made of Paracelsus's *ludus* by the liquor alkahest;[i] or, as Cardan relates, that an empiric had in his time, who travelled up and down Italy, curing those wherever he came that were tormented with the stone of the bladder[j] – if, I say, there be any such medicines, the physician may by such instruments perform that which, for ought appears, is not to be done by nature herself, since we never find that she dissolves a confirmed stone in the bladder. Nay, sometimes the physician does, even without the help of a medicine, control and overrule nature to the great and sudden advantage of the patient. For when a person, otherwise not very weak, happens by a fright or some surprising ill news to be so discomposed that the spirits hastily and disorderly thronging to some inward part, especially the heart, hinder

---

[6] Hippocrates, *Epidemics*, vi. 5. 1.

[h] Lat. lacks the rest of this sentence, and part of the next, resuming with 'in some cases . . .'.

[i] The term 'alkahest' was first used by Paracelsus, but developed by Helmont, who considered it to be the universal solvent. *Ludus*, according to Paracelsus, was a mineral compound capable of curing stone. According to Helmont, when dissolved in the alkahest, *ludus* yields an oily medicine for stone, *De lithiasi* (Amsterdam, 1648), vol. 2, pp. 62–3 (ch. 7, sections 22 & 23).

[j] Citing a work by the Italian physician and natural philosopher Girolamo Cardano (1501–76), probably *De malo recentiorum medicorum usu libellus* (Venice, 1536).

the regular and wonted motion of it, by which disorder the circulation of the blood is hindered or made very imperfect; in this case, I say, the patient is by nature's great care of the heart (as is commonly supposed even by physicians) cast into a swoon, whence the physician sometimes quickly frees him by rubbing and pinching the limbs, the ears and the nose, that the spirits may be speedily brought to the external parts of the body, which must be done by a motion to the circumference (as they call it), quite opposite to that towards the centre or heart which nature had given them before. But as to the theory of swoonings, I shall not now examine its truth, it being sufficient to warrant my drawing from thence an argument *ad hominem* that the theory is made use of by those I reason with.

By what has been discoursed one may perceive that, as there are some phenomena that seem to favour the doctrine of the naturists about the cure of diseases, so there are others that appear more manifestly favourable to the hypothesis we propose. And both these sorts of phenomena, being considered together, may well suggest a suspicion, that the most wise and yet most free author of things, having framed the first individuals of mankind so as to be fit to last many years, and endowed those protoplasts with the power of propagating their species, it thereupon comes to pass that in the subsequent hydraulico-pneumatical engines we call human bodies, when neither particular providence, nor the rational soul, nor overruling impediments interpose, things are generally performed according to mechanical laws and courses, whether the effects and events of these prove to be conducive to the welfare of the engine itself, or else cherish and foment extraneous bodies or causes whose preservation and prospering are hurtful to it. On which supposition it may be said that the happy things referred to nature's prudent care of the recovery and welfare of sick persons, are usually genuine consequences of the mechanism of the world and the patient's body, which effects luckily happen to be coincident with his recovery, rather than to have been purposely and wisely produced in order to it; since I observe that nature seems to be careful to produce, preserve and cherish things hurtful to the body as well as things beneficial to it. For we see in the stone of the kidneys and bladder, that out of vegetable or animal substances of a slighter texture – such as are the alimental juices which in sucking children (who are observed to be frequently subject to the stone in the bladder) are afforded by so mild a liquor as milk – nature skilfully frames a hard body of so firm a texture, that it puzzles physicians and chemists to tell how such a coagulation can be made of such substances: and I have found more than one calculus to resist both spirit of salt, that readily dissolves iron and steel, and that highly corrosive menstruum, oil of vitriol itself. We see also that divers times the seeds or seminal principles of worms that lie concealed in unwholesome fruits and other ill-qualified aliments are preserved and cherished in the body, so as, in spite of the menstruums, ferments, etc., they meet with there, they grow to be perfect worms (of their respective kinds) that are often very troublesome and sometimes very dangerous to the body that harbours them, producing, though perhaps not immediately, both more and more various distempers (especially here in England) than every physician is aware of. This reflection may very well be applied to those instances we meet with in good authors[7] of frogs and even toads whose spawn, being taken in with corrupted water, has been cherished in the stomach until, the eggs being grown to be complete animals, they produced horrid symptoms in the body that had lodged and fed them. And if according to the received opinion of physicians, stubborn quartans are produced by a melancholy humour seated in the spleen, it may be said that nature seems to busy herself to convert some parts of the fluid chyle into so tenacious and hardly dissipable a juice that in many patients, notwithstanding the neighbourhood of the spleen and stomach, neither strong emetics, nor purges, nor other usual remedies, are able in a long time to dislodge it or resolve it or correct it. But that is yet more conducive to my present purpose that is afforded me by the consideration of the poison of a mad dog, which nature sometimes seems industriously and solicitously to preserve. Since we have instances in approved authors that a little foam conveyed into the blood by a slight hurt (perhaps quickly healed up) is – notwithstanding the constant heat and perspirable frame of the human body and the dissipable texture of the foam – so preserved, and that sometimes for many years, that at the end of that long time, it breaks out and displays its fatal efficacy with as much vigour and fury as if it had but newly been received into the body.

To this agrees that which is well known in Italy about the biting of

---

[7] Schenck. Observ. Lib. 3. Pag. mihi 337 & seq. ['I find it at p. 337f.'.]

the tarantula. For though the quantity of poison can scarce be visible, since it is communicated by the tooth of so small an animal as a spider, yet in many patients it is preserved during a great part of their lives and manifests its continuance in the body by annual paroxysms. And I know a person of great quality, who complained to me that, being in the east, the biting or stinging of a creature whose offensive arms were so small that the eye could very hardly discern the hurt, had so lasting an effect upon him that, for about twelve years after, he was reminded of his mischance by a pain he felt in the hurt place about the same time of the year that the mischief was first done him. And in some hereditary diseases, as the gout, falling sickness and some kinds of madness, nature seems to act as if she did, with care as well as skill, transmit to the unhappy child such morbific seeds or impressions of the parent's disease, that in spite of all the various alterations the younger body passes through during the course of many years, this constantly protected enemy is able to exert its power and malice after forty or perhaps fifty years' concealment.

Such reflections as these, to which may be added, that the naturists make no scruple to style that death which men are brought to by diseases, a 'natural' death, make me backward to admit the famed sentence of Hippocrates hitherto considered, *morborum naturæ medici* [natures are the curers of diseases], without limitations – especially those two that are delivered in the fifth section,[8] to which I refer you the rather because they may help you to discern that divers phenomena that favour not the received notion of a kind and prudent being as nature is thought to be, are yet very consistent with divine providence.

---

[8] See Pag. 164. to Pag. 173. [Pp. 73–6 above.]

SECTION VIII

I have now gone through so many of the celebrated axioms concerning nature, that I hope I may reasonably presume that the other sentences of this kind (that my haste makes me leave unmentioned) will be thought capable of being fairly explicated – and with congruity to our hypothesis – by the help of the grounds already laid, since with light variations they may be easily enough improved and applied to those other particulars to which they are the most analogous.

But this intimation ought not to hinder me to make a reflection that not only is pertinent to this place, but which I desire may have retrospect upon a great part of the whole precedent discourse. And it is this: that though we could not intelligibly explicate all the particular axioms about nature and the phenomena of inanimate bodies that are thought (but not by me granted) to favour them by mechanical principles, it would not follow that we must therefore yield up the whole cause to the naturists. For we have already shown, and may do so yet further ere long, that the supposition of such a being as they call nature is far from enabling her partisans to give intelligible accounts of these and other phenomena of the universe. And though our doctrine should be granted to be, as well as that generally received about nature, insufficient to give good accounts of things corporeal, yet I shall have this advantage in this case: that a less degree of probability may serve in arguments employed but to justify a doubt, than is required in those that are to demonstrate an assertion.

It is true that the naturists tell us that the nature they assert is the principle of all motions and operations in bodies, which infers that in explicating them, we must have recourse to her. But before we acquiesce in, or confidently employ, this principle, it were very fit we knew what it is. This question I have discoursed of in the [fifth]^a section, but having there intimated a reference to another place, the importance as well as difficulty of the subject invites me to resume in this place the consideration of it, and both vary and add to what I formerly noted, that I may as well inculcate as clear my thoughts about it. I demand, then, of

---

those that assert such a nature as is vulgarly described, whether it be a substance or an accident. If it be the latter, it should be declared what kind of accident it is; how a solitary accident can have right to all those attributes, and can produce those numerous, manifold and wonderful effects that they ascribe to nature; and why a complex of such accidents as are the mechanical affections of matter (as figure, bulk, motion, etc.) may not altogether, as probably as that accident they call nature, be conceived to have been instituted by the perfectly wise author of the universe to produce those changes among bodies which are (at least for the most part) intelligibly referable to them? And if things be not brought to pass by their intervention, it were very fit, as well as desirable, that we should be informed by what other particular and intelligible means nature can effect them better than they may be by that complex.

But if it be said, as by most it is, that the principle called nature is a substance, I shall next demand whether it be a corporeal or an immaterial one. If it be said to be an immaterial substance, I shall further ask whether it be a created one or not. If it be not, then we have God under another name, and our dispute is at an end by the removal of its object or subject, which is said by the schools to be God's vicegerent, not God himself. But if nature be affirmed (as she is, at least by all Christian philosophers) to be a created being, I then demand whether or no she be endowed with understanding, so as to know what she does, and for what ends, and by what laws she ought to act. If the answer be negative, the supposition of nature will be of very little use to afford an intelligible account of things, an unintelligent nature being liable to the objections that will (a little below) be met with against the usefulness of nature, in case she be supposed a corporeal being. And though it should be said that nature is endowed with understanding and performs such functions as divers of the ancients ascribe to the soul of the world – besides that this hypothesis is near of kin to heathenism – I do not think that they who shall, with many Grecian and other philosophers who preceded Christianity, suppose a kind of soul of the universe, will find this principle sufficient to explicate the phenomena of it. For if we may compare the macrocosm and microcosm in this, as well as many are wont to do in other things, we may conceive that though nature be admitted to be endowed with reason, yet a multitude of phenomena may be mechanically produced without her immediate intervention; as we see that in man, though the rational soul has so narrow a province to take care of as the human body and is supposed to be intimately united to all the parts of it, yet abundance of things are done in the body by the mechanism of it without being produced by that soul. Of this we may allege as an instance, that in sleep, the circulation of the blood, the regular beating of the heart, digestion, nutrition, respiration, etc., are performed without the immediate agency or so much as the actual knowledge of the mind. And when a man is awake, many things are done in his body, not only without the direction, but against the bent of his mind, as often happens in cramps and other convulsions, coughing, yawnings, etc.

Nay, though some brutes, as particularly apes, have the structure of many parts of their bodies very like that of the analogous ones of human bodies, yet that admirable work of the formation and organisation of the foetus or little animal in the womb is granted by philosophers to be made by the soul of the brute (that is therefore said to be the architect of his own mansion), which yet is neither an incorporeal nor a rational substance. And even in a human foetus – if we will admit the general opinion of philosophers, physicians, divines and lawyers – I may be allowed to observe that the human body, as exquisite an engine as it is justly esteemed, is formed without the intervention of the rational soul, which is not infused into the body until this has obtained an organisation that fits it to receive such a guest – which is commonly reputed to happen about the end of the sixth week or before that of the seventh. And this consideration leads me a little further, and prompts me to ask how much, by the supposition or knowledge of the mind (at the newly mentioned time), we are enabled to explicate the manner how the forementioned functions of an embryo are performed, when at the end of six or seven weeks the rational soul supervenes and comes to be united to this living engine.

And if it be urged that nature being the principle of motion in bodies, their various motions (at least), which amount to a considerable part of their phenomena, must be explained by having recourse to her; I answer that it is very difficult to conceive how a created substance that is immaterial can, by a physical power or action, move a body, the agent having no impenetrable part wherewith to impel the corporeal mobile. I know that God, who is an immaterial spirit, ought to be acknowledged the primary cause of motion in matter, because (as we may justly, with

Monsieur Descartes,[b] infer) motion not belonging to corporeal substance as such, this must owe that to an incorporeal one. But then I
consider that there is that infinite distance between the incomprehen-
sible creator and the least imperfect order of his creatures, that we
ought to be very cautious how we make parallels between him and them
and draw inferences from his power and manner of acting to theirs –
since he, for instance, can immediately act upon human souls, as having
created them, but they are not able so to act upon one another.

And I think it the more difficult to conceive and admit that, if nature
be an incorporeal substance, she should be the greater mover of the
mundane matter – because we see that in a human body the rational
soul (which the school philosophers assert to be an immaterial spirit),
though vitally united to it, can only determine the motion of some of the
parts, but not give motion to any, or so much as regulate it in most. And
if nature be said to move bodies in another than a physical way, I doubt
whether the supposition of such a principle will be of much use to
physiologers in explicating phenomena, since I shall scarce think him an
inquisitive or a judicious doctor, who should imagine that he explains
that it gives an intelligible and particular account of the astonishing
symptoms of those strange diseases that divers very learned and sober
physicians impute to witchcraft, when he says that those strange
distortions and convulsive motions (for instance) and other prodigious
effects were produced by a wicked immaterial spirit called a devil. But
having to this purpose said more in another paper, which you may
command the sight of, I shall not trouble you with it here.[c]

The past discourse opposes their opinion, who assert nature to be an
immaterial creature. But because it is thought that a greater number of
philosophers, at least among the moderns, take her to be corporeal, I
shall now address my discourse to their hypothesis. And though I might
object that, if nature be a body, it may be demanded how she can
produce in men rational souls that are immaterial beings, and not
capable to be produced by any subtiliation or other change of matter
whatsoever; yet waiving this objection, I shall first demand whether
those I reason with believe nature, though corporeal, to act knowingly –
i.e. with consciousness of what she does and for predesigned ends – or
else to be blindly and necessarily moved and directed by a superior

---

[b]   See the first several articles in part 2 of *Principles of Philosophy*.
[c]   We do not know which work Boyle meant.

agent, endowed with (what she wants) an excellent understanding. And then I shall represent a few things, appliable some to one or the other of the two answers that may be made, and some to both.

And first, the Cartesians would ask how, if nature be a corporeal substance, we can conceive her capable of thinking; and, which is more, of being a most wise and provident director of all the motions that are made in the corporeal world.

Secondly, a philosophiser may justly ask how a corporeal being can so pervade and, as it were, compenetrate the universe, as to be intimately present with all its minute parts, whereof yet it is said to be the principle of motion.

Thirdly, he may also demand whence nature, being a material substance, comes itself to have motion, whereof it is said to be the principle, since motion does not belong to matter in itself, and a body is as truly a body when it rests as when it moves. And if it be answered that the first cause – that is, God – did at first put it into motion, I reply that the same cause may at least as probably be supposed to have put the unquestioned mundane matter into motion without the intervention of another corporeal being, in whose conception (i.e. as it is matter) motion is not involved.

Fourthly, it may likewise be asked how the laws of motion come to be observed or maintained by a corporeal being – which, as merely such, is either uncapable of understanding them or of acting with respect to them, or at least is not necessarily endowed with any knowledge of them or power to conform to them and to make all the parts of the unquestioned mundane matter do so too.

Fifthly, and I do not see how the taking in such an unintelligent and undesigning principle will free our understandings from great difficulties when we come to explicate the phenomena of bodies. For, as is elsewhere noted,[d] if nature be a bodily creature and acts necessarily and (if I may so speak) fatally, I see no cause to look upon it but as a kind of engine. And the difficulty may be as great, to conceive how all the several parts of this supposed engine called nature are themselves framed and moved by the great author of things, and how they act upon one another as well as upon the undoubted mundane bodies; as it is to conceive how, in the world itself – which is manifestly an admirably contrived automaton – the phenomena may, by the same author (who

---

d   This is discussed pp. 92–8.

was able to endow bodies themselves with active powers, as well as he could, on other scores, make them causes),[e] be produced by virtue and in consequence of the primitive construction and motions that he gave it (and still maintains in it), without the intervention of such a thing as they call nature – for this as well as the world being a corporeal creature, we cannot conceive that either of them act otherwise than mechanically. And it seems very suitable to the divine wisdom that is so excellently displayed in the fabric and conduct of the universe to employ in the world, already framed and completed, the fewest and most simple means by which the phenomena designed to be exhibited in the world could be produced. Nor need we be much moved by hearing some naturists say that nature, though not an incorporeal being, is of an order superior to mere matter, as divers of the schoolmen teach the things they call material forms to be. For who can clearly conceive an order or kind of beings that shall be real substances and yet neither corporeal nor immaterial? Nor do I see how the supposition of this unintelligible, or at least unintelligent being, though we should grant it to have a kind of life or soul, will much assist us to explicate the phenomena; as, if a man be acquainted with the construction of mills, he may as well conceive how corn is ground by a mill driven by the wind or by a stream of water, which are brute and senseless beings, as he can by knowing that it is kept at work by a horse who, though an animated being, acts in our case but as a part of an engine that is determined to go round, and who does neither intend to grind the corn nor know that he grinds it.

And in this place (though perhaps not the very fittest), I may question with what congruity to their master's doctrine the school philosophers teach that nature is the principle of motion in all the bodies they call natural. For – not to urge, that those great masses of sublunary matter to which they give the name of elements, and the mixed bodies that consist of them, are by divers learned men said to be moved to or from the centre of the earth by distinct internal principles which they call gravity in the earth and water, and levity in the fire and air; and that there is ascribed also to every compounded body that quality of the two, which belongs to the element that predominates in it – not to urge this, I say, consider that the celestial part of the world does so far exceed the subcelestial in vastness, that there is scarce any comparison between

[e]   Lat. lacks this parenthetical phrase, and the one following.

them. And yet the generality of the Peripatetics after Aristotle tell us that the celestial globes of light and the vast orbs they suppose them to be fixed in are moved from west to east by intelligences – that is, rational and separate beings, without whose conduct they presume that the motions of the heavens could not be so regular and durable as we see they are. So that, in that part of the universe which is incomparably vaster than the sublunary is, intelligences being the causes of motion, there is no recourse to be had to nature as the true and internal principle of it.

And here it may not perhaps be improper to declare somewhat more fully a point already touched upon – namely, that if to know what is the general efficient cause of motion can contribute to the explication of particular phenomena, the hypothesis of those naturists I now reason with will have no considerable advantage (if any at all) of ours, which derives them from the primitive impulse given by God to matter and from the mechanical affections of the greater and lesser portions of it. For it is all one to him that would declare by what particular motion (as swift, slow, uniform, accelerated, direct, circulate, parabolical, etc.) this or that phenomenon is produced, to know whether the motions of the parts of matter were originally impressed on them by nature or immediately by God – unless it be that he, being of infinitely perfect knowledge, may be, more probably than a creature, supposed to have at first produced in matter motions best accommodated to the phenomena that were to be exhibited in the world.

Nor do I see sufficient cause to grant that nature herself (whatever she be)^f produces any motion *de novo* [which is new], but only that she transfers and regulates that which was communicated to matter at the beginning of things. (As we formerly noted that, in the human body, the rational soul or mind has no power to make new motions, but only to direct those of the spirits and of the grosser organs and instruments of voluntary motion.) For – besides that many of the modern naturists approve of the Cartesian opinion that the same quantity of motion is always preserved in the whole mass of the mundane matter that was communicated to it at first, though it be perpetually transferring it from one part to another – besides this, I say, I consider that, if nature produces in these and those bodies motion that were never before in beings (unless much motion be annihilated, which is a thing as yet

---

^f Lat. lacks the parenthetical phrase.

unproved),[g] the quantity of motion in the universe must have for some thousands of years perpetually increased and must continue to do so – which is a concession that would much disorder the whole theory of local motion and much perplex philosophers, instead of assisting them, in explicating the phenomena of bodies.

And as for the effects of local motion in the parts of the universal matter – which effects make a great part of the phenomena of the world – after what I have formerly declared, you will not wonder to hear me confess that, to me, the supposition of nature – whether men will have her an immaterial or corporeal substance,[h] and either without knowledge or else endowed with understanding – does not seem absolutely necessary, nor perhaps very useful, to make us comprehend how they are produced. The bodies of animals are divers of them little less curiously framed than men's, and most of them more exquisitely than (for ought we know) the great inanimate mass of the corporeal world is. And yet, in the judgement of no mean naturalists, some of the mechanical philosophers that deny cogitation and even sense (properly so called) to beasts do – at least as intelligibly and plausibly as those that ascribe to them souls endowed with such faculties as make them scarce more than gradually different from human ones – explicate the phenomena that are observed in them. And I know not whether I may not on this occasion add that the Peripatetics themselves, especially the moderns, teach some things whence one may argue that the necessity of recurring to nature does not reach to so many things by far as is by them supposed. For the efformation (or framing) of the bodies of plants and animals, which are by great odds the finest pieces of workmanship to be met with among bodies, is ascribed not immediately to nature but to the soul itself, which they will have to be the author of the organisation of the body, and therefore call it the architect of its own mansion – which, they say, that it frames by an innate power and skill that some call plastic and to which others give other names. And unto the same soul, operating by her several functions, they attribute the concoction of aliments, the expulsion of excrements, the production of

---

[g] Lat. lacks the parenthetical phrase.

[h] In Lat. the paragraph begins: *Quod autem ad effectus motus localis in universalis materiae partibus attinet, quis forte crediderit Naturae hypothesim, sive eam velis materialem esse, aut immaterialem substantiam, ...* [But with regard to the effects of local motion on the parts of universal matter, whoever may have believed the hypothesis of nature – whether you want it to be a material or immaterial substance, ...]

milk, semen, etc., the appetitive, locomotive, and I know not how many other faculties ascribed to living bodies. And even in many inanimate ones, the noblest properties and operations are by the same school philosophers attributed to what they call their substantial forms, since from these they derive the wonderful properties of the loadstone, the attractive faculty of amber and other electrics, and the medical virtues of gems and other mineral bodies, whether consistent or fluid.

But not to insist on this argument, because it is but *ad hominem* (as they speak), if we consider the thing itself by a free examen of the pretended explanations that the vulgar philosophers are wont, by recurring to nature, to give of the phenomena of the universe, we shall not easily look on those accounts as meriting the name of explications. For to explicate a phenomenon, it is not enough to ascribe it to one general efficient, but we must intelligibly show the particular manner how that general cause produces the proposed effect. He must be a very dull enquirer who, demanding an account of the phenomena of a watch, shall rest satisfied with being told that it is an engine made by a watchmaker, though nothing be thereby declared of the structure and coaptation of the spring, wheels, balance and other parts of the engine; and the manner how they act on one another, so as to co-operate to make the needle point out the true hour of the day. And (to improve to my present purpose an example formerly touched upon)[i] as he that knows the structure and other mechanical affections of a watch will be able by them to explicate the phenomena of it, without supposing that it has a soul or life to be the internal principle of its motions or operations; so he that does not understand the mechanism of a watch will never be enabled to give a rational account of the operations of it by supposing – as those of China did when the Jesuits first brought watches thither – that a watch is an European animal, or living body, and endowed with a soul.

This comparison seems not ill to befit the occasion of propounding it, but to second it by another that is more purely physical: when a person unacquainted with the mathematics admires to see that the sun rises and sets in winter in some parts of the horizon, and in summer in others distant enough from them, that the day in the former season is by great odds shorter than in the latter, and sometimes (as some days before the middle of March and of September) the days are equal to the night; that

[i] Lat. lacks the parenthetical phrase.

the moon is sometimes seen in conjunction with the sun, and sometimes in opposition to him, and between those two states is every day variously illuminated;[j] and that sometimes one of those planets, and sometimes another, suffers an eclipse – this person, I say, will be much assisted to understand how these things are brought to pass, if he be taught the clear mathematical elements of astronomy. But, if he be of a temper to reject these explications as too defective, it is not like[ly] that it will satisfy him to tell him after Aristotle and the schoolmen that the orbs of the sun and moon and other celestial spheres are moved by angels or intelligences, since to refer him to such general and undetermined causes will little, or not at all, assist him to understand how the recited phenomena are produced.

If it be here objected that these examples are drawn from factitious, not from merely physical, bodies, I shall return this brief answer and desire that it be applied not only to the two freshly mentioned examples, but to all of the like kind that may be met with in this whole treatise (near the beginning of which, had I remembered it, something to the same purpose should have had place). I say then, in short, that divers of the instances we are speaking of are intended but for illustrations, and that others may be useful instances if they should be no more than analogous ones, since examples drawn from artificial bodies and things may have both the advantage of being more clearly conceived by ordinary understandings, and that of being less obnoxious to be questioned in that particular in which the comparison or correspondence consists. And I the less scruple to employ such examples, because Aristotle himself, and some of his more learned followers, make use of divers comparisons drawn from the figures and other accidents of artificial things to give an account of physical subjects, and even of the generation, corruption and forms of natural bodies.

This advertisement premised, I pursue the discourse it interrupted, by adding that thus we see that confirmed which was formerly observed – namely, that though mechanical principles could not be satisfactorily employed for explaining the phenomena of our world, we must not therefore necessarily recur to and acquiesce in that principle that men call nature, since neither will that intelligibly explain them. But in that case, we should ingeniously confess that we are yet at a loss how they are performed, and that this ignorance proceeds rather from the natural

---

[j] Lat. lacks 'and between those two ... variously illuminated'.

imperfection of our understandings than from our not preferring nature (in the vulgar notion of it) to the mechanical principles in the explication of the phenomena of the universe. For whereas Monsieur Descartes[k] and other acute men confidently teach that there are scarce any of these phenomena that have been truly and intelligibly deduced from the principles peculiar to the Aristotelians and school philosophers, it will scarce be denied by any that is acquainted with physico-mathematical disciplines, such as optics, astronomy, hydrostatics and mechanics more strictly so called, but that very many effects (whereof some have been handled in this present tract) are clearly explicable by mechanical principles – which, for that reason, Aristotle himself often employs in his *Quæstiones mechanicæ*[l] and elsewhere. So that if, because the corpuscularian principles cannot be satisfactorily made use of to account for all that happens among things corporeal, we must refuse to acquiesce in them, it is but just that since a recourse to what is called nature is yet more dark and insufficient, at least we must reject as well the latter as the former hypothesis, and endeavour to find some other preferable to both.

And now, if it be demanded what benefit may redound to a reader from the explications given in the foregoing seventh section and, in general, from the troublesome, as well as free, *Enquiry* whereof they make a considerable part; I shall answer that I am not quite out of hope that the things hitherto discoursed may do some services both to natural philosophy and to religion. And as to the first of these, this tract may be of use to the cultivators of that science by dissuading them from employing often and without great need in their philosophical dis-courses and writings a term (I mean 'nature') which, by reason of its great ambiguity and the little or no care which those that use it are wont to take to distinguish its different acceptions, occasions both a great deal of darkness and confusedness in what men say and write about things corporeal and a multitude of controversies – wherein really men do but wrangle about words, while they think they dispute of things, and perhaps would not differ at all if they had the skill or luck to express themselves clearly. Besides which service the past discourse may do this other: to wean many from the fond conceit they cherish, that they

---

[k] Probably a reference to a passage found only in the preface to the French edition (1647) of *Principles of Philosophy*, which was first published in Latin.
[l] Spuriously attributed to Aristotle.

understand or explicate a corporeal subject or a phenomenon when they ascribe it to nature. For to do that one needs not be a philosopher, since a country swain may easily do the same thing.

On this occasion, I must not forbear to take notice that the unskilful use of terms of far less extent and importance, and also less ambiguous, than the word 'nature' is, has been and still is, no small impediment to the progress of sound philosophy. For not only the greatest part both of physicians (though otherwise learned men) and of chemists, but the generality of physiologers too have thought that they have done their part, though not on all occasions yet on very many, when they have referred an effect or a phenomenon to some such things as those that are presumed to be real qualities, or are by some styled natural powers, or are by others, by a more comprehensive and more usual name (which [I] therefore here chiefly employ) called faculties – for each of which they are wont to form a name fit for their purpose, though they do not intelligibly declare what this faculty is and in what manner the operations they ascribe to it are performed by it. Thus the attractive faculty ascribed to a man that is enabled by nature's (presumed) abhorrence of a vacuum to suck up drink through a straw or pipe has been for many ages acquiesced in as the true cause of the ascension of that liquor in suction – of which nevertheless the modern philosophers that have slighted explications derived merely from faculties have assigned (as has been already declared) intelligible and even mechanical causes. The power that a loadstone has with one pole to attract (as they speak) the northern point of the mariner's needle and with the other to drive it away is looked upon as one of the noblest and most proper faculties of that admirable stone. And yet I elsewhere show[m] how in a very small indeed, but true and natural magnet, I have by a bare and sometimes invisible change of texture given that extreme of the magnet, that before drew the southern point of the needle, the power to draw the northern, and to the opposite extreme, the power to drive it away; so much does even this wonderful attractive faculty, as it is called, depend upon the mechanical structure of the mineral and its relation to other bodies among which it is placed, especially the globe of the earth and its magnetical effluvia.

But because in another paper I purposely discourse of what naturists

---

[m] See experiment IV in 'Experiments and Notes about the Mechanical Production of Magnetism', *Works*, vol. 4, pp. 341f.

call faculties,[n] I shall here content myself to note in general that the term 'faculty' may indeed be allowed of, if it be applied as a compendious form of speech but not as denoting a real and distinct agent, since in reality the power or faculty of a thing is (at least) oftentimes but the matter of it made operative by some of its mechanical modifications. [I say 'some', because the complex of all makes up its particular nature.][o] And with how little scruple soever men commonly speak of faculties, as supposing them to be distinct and active principles, yet this condition does not necessarily belong to them. For sometimes, if not frequently, the effect of what is reputed a natural power or faculty is produced by the texture, figure and, in a word, mechanical disposition of the agent, whereby it determines the action of a remoter agent to the produced effect. Thus in a clock, to make the balance vibrate, to point at the hour, to make at set times the hammer strike upon the bell, are but different effects of the weight or spring that sets and keeps the engine in motion. And so a key may either acquire or lose its power of opening a door (which perhaps some schoolmen would call its aperitive faculty) by a change, not made in itself, but in the locks it is applied to or in the motion of the hand that manages it.

And lest it should be objected that these instances are taken wholly from artificial bodies, I shall add that when a clear piece of native crystal has obtained, as it often does, a good prismatical shape and is in a due position exposed to the sunbeams, its figuration, by enabling it to refract and reflect those beams after a certain manner, gives it a colorific faculty whereby it is enabled to exhibit that wonderful and pleasing variety of colours that emulate if not surpass those of the rainbow. And so in a concave metalline looking-glass, though there seem to be many distinct faculties, such as that of reflecting, inverting, magnifying divers objects, and melting, burning, etc. several bodies, yet all these powers are but the genuine consequences of the figure, capacity and smoothness, which are mechanical affections of the matter of the speculum. And, indeed, if I judge aright (though what I am going to say will seem a paradox), yet many qualities of very many bodies are but lasting dispositions to be thus or thus wrought upon by the action of external agents, and also (perchance) to modify that action; as we see that the power of making an echo that is observed in divers hollow places is nothing but the

---

[n] Probably a reference to *The Origin of Forms and Qualities*.
[o] The square brackets are Boyle's.

mechanical disposition their figure and resistance gives them to reflect a sound. And, to resume the lately mentioned instance of a key, we may add that, by bare position, either end of it, especially if the key be long, may be made to acquire or lose a transient magnetic faculty from the effluvia of that great magnet, the earth; and that also the same key may, in a few moments, acquire a durable magnetism by a mechanical change received from the loadstone, as is known to those that are anything versed in the philosophy of that wonderful mineral.

And to me it seems likely, that one main reason why learned men have ascribed such inherent and active powers as they call faculties to so many bodies is because that, not being conversant enough with natural and artificial things, they did not duly perpend how great a difference there may be between a body considered absolutely (or by itself) and the same body considered in such circumstances as it may be found in. For in some cases a physical body may have strange things justly ascribed to it, though not as it is such a body considered simply or unassociated with other bodies, but as it is placed among congruous ones and makes the principal or most operative part of a compounded body or of the complex of bodies it is joined with, and which are of such determinate structures as are convenient for the phenomena to be exhibited.

This may be analogically seen in what happens to a spring. For if, being bent, it is held in one's hand or crowded into a box, it is but a simple thing that does only, by its expansive endeavour, strive to remove the bodies that keep it compressed. But in a curious watch, it may, by virtue of the structure of that engine, become the principle of I know not how many differing and perhaps contrary motions among the parts of it, and of many notable phenomena and effects exhibited or produced thereby. This reflection may perhaps be improved if I here add that in many bodies a fluid substance, determined to convenient motions, may be equivalent to an internal spring, especially if it be assisted by friendly external agents. This may be illustrated by considering that, if one that plays skilfully on a flute blow[s] out of his mouth into the open air, he will but turn it into a vapid aerial stream; but if this wind duly pass into the instrument and be modified there by the musician's fingers and skill, the simple stream of air may be formed into very various and melodious tunes. Thus gunpowder artificially tempered, though if it be fired in the open air, it will give only a rude and sudden flash that presently vanishes, yet if it be skilfully disposed of in rockets and other well-contrived instruments and then kindled, it will exhibit a great and pleasing variety of shining bodies and phenomena that are justly admired in the best sort of artificial fireworks. A physical instance also in favour of our analogical or vicarious springs (if I may so call them) is afforded me by the bulbs of onions and the roots of aloes (commonly called *semper-vive* [always alive]) and some other vegetables, which in the spring being exposed to the air, the juices and spirits contained in them will be so agitated by the warmth of that season and so modified by the particular structure of the more firm parts that, though neither earth nor rain co-operate, they will shoot forth green stalks or leaves for many weeks together, as if they were planted in a good soil (though the matter of these green productions be furnished by the radical parts themselves, as may be argued both from the manifest diminution of the bulb in bigness and the great and gradual decrement in weight that I observed in making experiments of this kind).[P] And so also the air, which is an external fluid, concurring with the juices and spirits of divers insects and other cold animals, may both be put into motion and have that motion so determined by their organisation as to recover in the spring or summer (as it were) a new life, after they have lain moveless and like dead things all the winter; as we see in flies that in a hot air quickly recover motion and sense after having lost both for perhaps many months. And the like change may be far more suddenly observed in them in the warmer seasons of the year, when the air is drawn from them by the pneumatic pump and afterwards permitted to enliven them again. And to give another instance that may possibly please better (because, as it is purely physical, so it is simple and very conspicuous): though that which the sunbeams are wont primarily to produce be but light and perhaps heat, yet falling in a due manner upon a rorid cloud, they form there the figure of a vast bow and, being variously reflected and refracted, adorn it with the several colours men admire in the rainbow.

But I must not farther prosecute an observation that I mentioned but occasionally, as an instance whereby to show that the advancement of solid philosophy may be much hindered by men's custom of assigning as true causes of physical effects imaginary things or perhaps arbitrary names, among which none seems to have had a more malevolent influence upon physiology than the term 'nature', none having been so

[P]   The closing parenthesis is missing in the first edition.

frequently and confidently used or employed to so many differing purposes. And therefore, though I would not totally forbid the use of the word 'nature', nor of expressions of kin to it, in popular discourses or even in some philosophical ones where accurateness is not required or ambiguity is prevented by the context, nor (to dispatch) where it may be employed as a compendious form of speech without danger to truth or prejudice to sound philosophy (in which cases I myself forbear not the use of it); yet I hope our *Free Enquiry* may (somewhat at least) conduce to the more skilful indagation and happy discovery of physical truths, if it can persuade men to make use less frequently and with more circumspection of so ambiguous and so often abused a term as 'nature', and cease to presume that a man has well performed the part of a true physiologer till he have circumstantially or particularly deduced the phenomenon he considers by intelligible ways from intelligible principles – which he will be constantly put in mind of doing, or discover that he has not done it, if by forbearing general and ambiguous terms and words, he endeavours to explain things by expressions that are clear to all attentive readers, furnished with an ordinary measure of understanding and reason. And this perspicuous way of philosophising should not be a little recommended to ingenious men, by the valuable discoveries which those that have employed it in their researches and explications of difficult things have in this inquisitive age happily made, not only about the various phenomena commonly referred to the *fuga vacui* [avoidance of a vacuum], but in the hydrostatics, optics, anatomy, botanics and divers other parts of real learning that I cannot now stay to enumerate. And thus much it may possibly be sufficient to have said about the service our doctrine may do natural philosophy.

As for religion, if what I have formerly said in favour of it be duly considered and applied, the past discourse will not appear unfriendly nor perhaps useless to it. And therefore, if I do here abridge what I have there said and add to it some considerations that were fit to be reserved for this place, I hope the doctrine we have proposed may appear fit to do it a threefold service.

I.   And in the first place, our doctrine may keep many that were wont, or are inclined, to have an excessive veneration for what they call nature from running, or being seduced, into those extravagant and sacrilegious errors that have been upon plausible pretences embraced not only by many of the old heathen philosophers, but by divers modern professors of Christianity who have of late revived, under new names and dresses, the impious errors of the Gentiles. This I venture to say, because many of the heathen writers (as has been shown in the fourth section) acknowledged indeed a God (as these also own they do), but meant such a god as they often too little discriminated from matter, and even from the world, and as is very differing from the true one adored by the Christians and Jews.[1] For ours is a God, first, infinitely perfect; and then secondly, by consequence, both incorporeal and too excellent to be so united to matter as to animate it like the heathen's mundane soul, or to become to any body a soul properly so called; and thirdly, uncapable of being divided and having either human souls or other beings (as it were) torn or carved out or otherwise separated from him, so as to be truly parts or portions of his own substance. Whereas the idolaters and infidels I speak of conceived, under the name of God, a being about which they dogmatically entertained conceptions which, though different from one another, are much more so from the truth.[2]

[1] Thus the Stoics, in Laertius, describe the world thus: *Mundus est qui constat ex Coelo et Terra atque ex illorum naturis; sive, Qui constat ex Diis & hominibus, iisque rebus quæ horum gratia conditæ sunt.* [The world is what is made up of heaven and earth and their natures; or again, what is made up of gods and men and the things constituted for the latter's sake.] And of Chrysippus, one of the patriarchs of that sect, the same historian in the same book (Diog. Laertius, book 7, in *Vita Zenonis*) says, *Purissimum dixit ac liquidissimum æthera, quem etiam primum asserunt Stoici esse Deum, sensibiliter veluti infusum esse, per ea quæ sunt in aere, per cunctas animantes et arbores, per terram autem ipsam secundum halitum.* [From the life of Zeno of Citium in Laertius's *De vitis philosophorum*. The first passage, *Zeno* 138, is itself a quotation from the lost *De phaenomenis caelestibus* by the Stoic Posidonius of Apamea (*c.*135–50 BC). The second passage (ibid. 139) is from *De providentia* by Chrysippus (*c.* 279–206 BC), likewise a Stoic: 'He said it is very pure and very liquid ether, which the Stoics declare to be the foremost god, and sensibly to have pervaded, like breath, all that is in the air, all animals and plants, and the whole earth itself.'] To which agrees not only that noted passage of Virgil, *Principio coelum,* etc., but another, which I somewhat wonder learned men should read with no more reflection, since he there gives the sky the very title of the high God: *Tum Pater Omnipotens foecundis imbribus æther,* &c. [Virgil, *Georgics*, ii.325: 'then the almighty father, Ether, with fertile rains'.]

[2] The error here rejected was the opinion of many of the heathen philosophers, and particularly of the Stoical sect, of whose author, Laertius (in *Vita Zenonis*) says, *De divina substantia Zeno ait mundum totum atque coelum.* [On the substance of God, Zeno says it is the whole world and heaven. (*Zeno* 148)] And several ethnic philosophers, even after the light of the gospel began to shine in the world, adopted the argument of the elder Stoics, who inferred the world to be animated and rational from the nature of the human soul, which they thought a portion of the intelligent part of the world that some of them confounded with the deity. For the Stoics (in Laertius, *De vitis phil.*, book 7) affirm, *Mundum esse animale et rationale et animatum* (ἔμψυχον) *et intelligibile* [that the world is a living, rational, animate, intelligent thing (ibid., 142)]. And it is added, *mundum animatum esse, inde manifestum est, quod anima nostra inde veluti avulsa sit.* [The world is animate, in that our souls are clearly wrought from it. (ibid., 143)] Thus Seneca (*Epist.* 92.), *Quid est autem cur non existimes, in eo divini aliquid existere, quæ dei pars est?* [From Seneca, *Epistulae morales*: What is your reason for not believing there to be something divine in that which is a part of God?] So Plutarch, speaking of the soul (in *Quæstiones Plut.*), *Non opus solum*

For first, most of them thought their God to be purely corporeal, as – besides what Diogenes Laertius and others relate – I remember Origen does in several places affirm. If you will believe Eusebius,[3] the ancient Egyptian theologers not only affirmed the sun, moon and stars to be gods, but denied incorporeal substances or invisible natures to have framed the world, but only the sun that is discoverable to our eyes. And this corporiety of God seems manifestly to be the opinion of Mr Hobbes and his genuine disciples,[q] to divers of whose principles and dogmas it is as congruous as it is repugnant to religion. But secondly, there are others that allowed a soul of the world, which was a rational and provident being, together with the corporeal part of the universe, especially heaven – which, I remember, Aristotle himself styles a divine body[4] (or, as some render his expressions, the body of God) – but withal they held that this being did properly inform this great mass of the universe and so was, indeed, a mundane soul. And though some of our late infidels (formerly pointed at in this treatise) pretend to be great discoverers of new light in this affair, yet as far as I am informed of their doctrine, it has much affinity with, and is little or not at all better than, that which I formerly noted out of Lactantius to have been asserted by the Stoics, and the doctrine which is expressed by Maximus (a pagan) to St Austin.[r] *Equidem unicum esse deum summum atque magnificum quis tam demens, tam mente captus, ut neget esse certissimum? Hujus nos virtutes per mundanum opus diffusas, multis vocabulis invocamus, quoniam nomen ejus cuncti proprium ignoremus.* [Now who would be so foolish, so bereft

---

dei, sed et pars est; neque ab ipso, sed ex ipso nata est. [Plutarch, *Quaestiones naturales*: 'It is not the work of God, but a part of him: it was not created by him, but born of him.'] And Epictetus (*Dissert.* I. cap. 14.), *Animæ ita alligatæ et conjunctæ deo sunt, ut particulæ eius sunt.* [From the *Discourses* by the the Stoic philosopher Epictetus (*c.* 50–*c.* 130): Souls are so closely bound and conjoined to God that they are a part of him.]

3   Præpar. lib. 3. cap. 4. [*Praeparatio gospelis*, by Eusebius (260–340), bishop of Caesarea.]
4   De Coelo l. 2. c. 3. [Aristotle, *De caelo*.]

q   Hobbes believed that God is corporeal, which led Boyle and many others to think him an atheist.
r   Citing a letter to Augustine from Maximus, an otherwise unknown person who was probably his former teacher. Boyle edits the passage, which reads in full: *Et quidem unum esse deum summum sine initio, sine prole naturae ceu patrem magnum atque magnificum quis tam demens, tam mente captus neget esse certissimum? Huius nos virtutes per mundanum opus diffusas, multis vocabulis invocamus, quoniam nomen eius cuncti proprium videlicet ignoramus.* [Now who would be so foolish, so bereft of their wits, as to deny the certainty of there being a single, supreme and glorious God, without beginning, without natural offspring, a great and magnificent father? His virtues, spread throughout his worldly creation, we refer to under a multitude of names, for his own name is of course unknown to us all.] Augustine, *Letters*, xvi.

of their wits, as to deny the certainty of there being a single, supreme and glorious God? His virtues, spread throughout his worldly creation, we refer to under a multitude of names, for his own name is unknown to us all.] Or by that famous and learned Roman, Varro, who is cited by St Austin to have said, *Deum se arbitrari animam mundi, et hunc ipsum mundum esse Deum: sed sicut hominem sapientem, cum sit ex animo et corpore, tamen ab animo dicimus sapientem; ita mundum deum dici ab animo, cum sit ex animo et corpore.* [He (Varro) considered God to be the soul of the world, and the world itself to be God. But just as a wise man, though made up of mind and body, is called wise for his mind, so the world is called God for its mind, though it too is made up of mind and body.][5]

[II.] The doctrine by us proposed may (it is hoped) much conduce to justify some remarkable proceedings of divine providence, against those cavillers that boldly censure it upon the account of some things that they judge to be physical irregularities (for moral ones concern not this discourse), such as monsters, earthquakes, floods, eruptions of volcanoes, famines, etc. For according to our doctrine:

1. God is a most free agent, and created the world not out of necessity but voluntarily, having framed it as he pleased and thought fit at the beginning of things, when there was no substance but himself and consequently no creature to which he could be obliged, or by which he could be limited.

2. God having an understanding infinitely superior to that of man in extent, clearness and other excellencies, he may rationally be supposed to have framed so great and admirable an automaton as the world, and the subordinate engines comprised in it, for several ends and purposes, some of them relating chiefly to his corporeal and others to his rational creatures; of which ends he has vouchsafed to make some discoverable by our dim reason, but others are probably not to be penetrated by it but lie concealed in the deep abyss of his unfathomable wisdom.

3. It seems not incongruous to conceive that this most excellent and glorious being thought fit to order things so that both his works and actions might bear some signatures and, as it were, badges of his attributes, and especially to stamp upon his corporeal works some

---

[5] De Civit Dei lib. 7. cap. 6. [Augustine, *The City of God*, citing the Roman antiquarian, philosopher and grammarian Marcus Varro (116–27 BC).]

tokens or impresses, discernible by human intellects, of his divine wisdom – an attribute that may advantageously disclose itself to us men by producing a vast multitude of things from as few and as simple principles, and in as uniform a way, as (with congruity to his other attributes) is possible.

4. According to this supposition, it seems that it became the divine author of the universe to give it such a structure and such powers, and to establish among its parts such general and constant laws, as best suited with his purposes in creating the world; and to give these catholic laws and particular parts or bodies such subordinations to one another, and such references to the original fabric of the grand system of the world, that on all particular occasions, the welfare of inferior or private portions of it should be only so far provided for as their welfare is consistent with the general laws settled by God in the universe, and with such of those ends that he proposed to himself in framing it, as are more considerable than the welfare of those particular creatures.

Upon these grounds, if we set aside the consideration of miracles as things supernatural, and of those instances wherein the providence of the great rector of the universe and human affairs is pleased peculiarly to interpose, it may be rationally said that God – having an infinite understanding to which all things are at once in a manner present – did, by virtue of it, clearly discern what would happen in consequence of the laws by him established in all the possible combinations of them and in all the junctures of circumstances wherein the creatures concerned in them may be found. And that, having – when all these things were in his prospect – settled among his corporeal works general and standing laws of motion suited to his most wise ends, it seems very congruous to his wisdom to prefer (unless in the newly excepted cases) catholic laws and higher ends before subordinate ones, and uniformity in his conduct before making changes in it according to every sort of particular emergencies; and consequently, not to recede from the general laws he at first most wisely established to comply with the appetites or the needs of particular creatures, or to prevent some seeming irregularities (such as earthquakes, floods, famines, etc.) incommodious to them – which are no other than such as he foresaw would happen (as the eclipses of the sun and moon from time to time, the falling of showers upon the sea and sandy deserts, and the like must do, by virtue of the original disposition of things) and thought fit to ordain or to permit, as not unsuitable to some or other of those wise ends which he may have in his all-pervading view – who, either as the maker and upholder of the universe or as the sovereign rector of his rational creatures, may have ends, whether physical, moral or political (if I may be allowed so to distinguish and name them), divers of which, for ought we can tell or should presume, are known only to himself. Whence we may argue, that several phenomena which seem to us anomalous may be very congruous or conducive to those secret ends, and therefore are unfit to be censured by us, dim-sighted mortals.

And indeed, the admirable wisdom and skill that, in some conspicuous instances, the divine opificer has displayed in the fitting of things for such ends and uses, for which (among other purposes) he may rationally be supposed to have designed them, may justly persuade us that his skill would not appear inferior in reference to the rest also of his corporeal works, if we could as well in these as in those discern their particular final causes. As if we suppose an excellent letter about several subjects and to different purposes, whereof some parts were written in plain characters, others in ciphers, besides a third sort of clauses wherein both kinds of writing were variously mixed, to be heedfully perused by a very intelligent person; if he finds that those passages that he can understand are excellently suited to the scopes that appear to be intended in them, it is rational as well as equitable in him to conclude that the passages or clauses of the third sort, if any of them seem to be insignificant or even to make an incongruous sense, do it but because of the illegible words; and that both these passages and those written altogether in ciphers would be found no less worthy of the excellent writer[6] than the plainest parts of the epistle, if the particular purposes they were designed for were as clearly discernible by the reader. And perhaps you will allow me to add that by this way of ordering things – so that in some of God's works the ends or uses may be manifest and the exquisite fitness of the means may be conspicuous [as the eye is manifestly made for seeing, and the parts it consists of admirably fitted to make it an excellent organ of vision],[s] and in others the ends designed seem to be beyond our reach – by this way (I say) of managing things,

---

[6]   See the Discourse of Final Causes. [*A Disquisition about the Final Causes*, esp. section IV, where Boyle discusses the degree to which we can know God's purposes; *Works*, vol. 5, pp. 420–44.]

[s]   The square brackets are Boyle's.

the most wise author of them does both gratify our understandings and make us sensible of the imperfection of them.

If the representation now made of providence serve (as I hope it may) to resolve some scruples about it, I know you will not think it useless to religion. And though I should miss of my aim in it, yet since I do not dogmatise in what I propose about it but freely submit my thoughts to better judgements, I hope my well-meant endeavours will be – as well as the unsuccessful ones of abler pens have been – excused by the scarce superable difficulty of the subject. However, what I have proposed about providence being written rather to do a service to theology than as necessary to justify a dissatisfaction with the received notion of nature that was grounded mainly upon philosophical objections, I hope our *Free Enquiry* may – though this second use of it should be quite laid aside – be thought not unserviceable to religion, since the first use of it, above delivered, does not depend on my notions about providence, no more than the third, which my prolixity about the former makes it fit I should in a few words dispatch.

III.   The last then, but not the least, service I hope our doctrine may do religion is that it may induce men to pay their admiration, their praises and their thanks, directly to God himself, who is the true and only creator of the sun, moon, earth and those other creatures that men are wont to call the works of nature. And in this way of expressing their veneration of the true God (who, in the holy scripture, styles himself a 'jealous God', Exodus 20:5) and their gratitude to him, they are warranted by the examples of the ancient people of God, the Israelites, and not only by the inspired persons of the Old Testament, but by the promulgators of the New Testament, and even by the celestial spirits, who, in the last book of it, are introduced praising and thanking God himself for his mundane works, without taking any notice of his pretended vicegerent, nature.[7]

The Conclusion

And now, dear Eleutherius, you have the whole bundle of those papers that I found and tacked together (for they are not all that I have written), touching my *Free Enquiry into the Received Notion of Nature*: at

---

[7]   Revelation 4:2.

the close of which essay, I must crave leave to represent two or three things about it.

1.   Since this treatise pretends to be but an *Enquiry*, I hope that any discourses or expressions that you may have found dogmatically delivered about questions of great moment or difficulty will be interpreted with congruity to the title and avowed scope of this treatise, and that so favourable a reader as Eleutherius will consider that it was very difficult in the heat of discourse never to forget the reserves that the title might suggest – especially since on divers occasions I could not have spoken with those reserves, without much enervating my discourse, and being by restrictions and other cautious expressions tedious or troublesome to you. But this, as I lately intimated, is to be understood of things of great moment or difficulty. For otherwise there are divers notions, suppositions and explanations in the vulgarly received doctrine of nature and her phenomena which I take to be either so precarious, or so unintelligible, or so incongruous, or so insufficient, that I scruple not to own that I am dissatisfied with them and reject them.

2.   Though upon a transient view of these papers, I find that several parcels that came first into my hands, having been laid and fastened together (to keep them from being lost, as others had already been) before the others were lighted on, some of them will not be met with in places that are the most proper for them, yet haste and sickness made me rather venture on your good nature for the pardon of a venial fault, than put myself to the trouble of altering the order of these papers and substituting new transitions and connections in the room of those with which I formerly made up the chasms and incoherency of the tract you now receive. And if the notions and reasonings be themselves solid, they will not need the assistance of an exact method to obtain the assent of so discerning a reader as they are presented to – upon the score of whose benignity, it is hoped that the former advertisement may likewise pass for an excuse, if the same things for substance be found more than once, in a tract written at very distant times and in differing circumstances. For, besides that such seeming repetitions will not (if I be not mistaken) frequently occur and will for the most part be found, by being variously expressed, to elucidate or strengthen the thought or argument they belong to; and besides that the novelty and difficulty of some points may have made it needful not only to display, but to inculcate them – besides these things, I say, it is very possible that the same notion may

serve to explicate or prove several truths and therefore may, without impertinency, be made use of in more than one part of our treatise. And if our *Enquiry* shall be thought worthy to be transcribed and presented to you a second time, after I shall have reviewed it and heard objections against it and considered the things that either you or I myself may find fault with in it, it is very possible that (if God grant me life and leisure) this tract – which, in its present dress, I desire you would look on but as an apparatus (towards a more full and orderly treatise) – may appear before you in a less unaccurate method, and that my second thoughts may prove more correct, more mature, or better backed and fortified than my first.

3. The subject of my enquiry being of great extent as well as consequence, it obliged me to consider and treat of many things (as philosophical, medical, theological, etc.) and, among them, of divers that are not at all of easy speculation. And I found it the more difficult to handle them well, because the attempt I have ventured upon being new and to be prosecuted by discourse. Many of them [being] opposite to the general sentiments of mankind, I was not to expect much assistance from anything but truth and reason. And therefore, as I cannot presume not to need your indulgence, so I cannot despair of obtaining it if in this my first essay, upon a variety of difficult points, I have not always hit the mark and as happily found the truth as sincerely sought it. But if you shall (which it is very probable you will) find that I have fallen into some errors, it will be but one trouble for you to make me discern them and forsake them (especially any wherein religion may be concerned), which I have by way of provision made it the more easy for myself to do, because (if my style have not wronged my intentions) I have written this discourse rather like a doubting seeker of truth than a man confident that he has found it.[t]

FINIS

---

[t] Here Lat. adds a quotation from Seneca, *Natural Questions*, i.25: *Veniet tempus quo posteri nos tam aperta nescisse mirentur.* [There will come a time when posterity will wonder that we were ignorant of things so manifest.]

Glossary

acception   the act of accepting

advertisement   notification, instruction

animadversive   percipient

aperitive   opening

architectonic   serving structural function

aspera arteria   trachea

chyle   white milky fluid formed in course of digestion

coaptation   adjustment of things to one another

connatural   inherent by nature or from birth

consecution   sequence

contemporate   moderate

coryza   catarrh

crasis   constitution

current   generally accepted

decrement   decrease

deglutition   the act of swallowing

depuration   the process of freeing from impurities

diadrom   the vibration of a pendulum

divers   several

effatum, effata   dictum, dicta

efformation   shaping

emetic   causing vomiting

empiric   medical practitioner lacking formal education

emunctory   a cleansing organ or canal

equipollent   of equal force

ethnic   heathen

examen   examination

flower-de-luce   the ornamented needle on the compass

fuliginous reeks   perspiration resulting from bodily heat

heresiarch   leader or founder of a heresy

heteroclite   irregular or unusual

imposthume   abscess or purulent swelling; cyst

incogitantly   without consideration, thoughtlessly

indagation   investigation

like   likely, probable

lily   the fleur-de-lis which marks the north on a compass

menstruum   a discharge or solvent

metastasis   transference or transformation

opacous   opaque

opificer   maker, framer

paraphrastical   of the nature of paraphrase

peccant   unhealthy, inducing disease or sickness

perpend   consider

phlebotomy   blood-letting

physiologer   student of nature

pica   a perverted craving for substances unfit for food, such as chalk

pregravitate   to gravitate more (than something else)

prepollent   predominating

preternatural   outside the ordinary course of nature

protoplast   archetype

quadrature(s)   the point(s) at which the moon is 90° distant from the sun

quartan   intermittent fever recurring at regular intervals

quiddity   the essence of a thing

ravings   uncontrolled thoughts

redargution   refutation

renitency   resistance

rorid   dewy

rumb(s)   the lines followed by vessels or winds moving continuously in one direction

seminal   containing the potential for future development

speculum   mirror

squib   a small fire-cracker

subtiliation   sublimation

sudorific   causing perspiration

sustentation   maintenance at a given level

tenent   tenet

turbith-mineral   basic sulphate of mercury

turgescence   swelling

us-ward   in our direction

uvea   the posterior coloured surface of the eye

vellicate   irritate

vicegerent   deputy ruler

wen   tumour

Index

Cambridge Texts in the History of Philosophy

Titles published in the series thus far

Antoine Arnauld and Pierre Nicole *Logic or the Art of Thinking* (edited by Jill Vance Buroker)

Boyle *A Free Enquiry into the Vulgarly Received Notion of Nature* (edited by Edward B. Davis and Michael Hunter)

Conway *The Principles of the Most Ancient and Modern Philosophy* (edited by Allison P. Coudert and Taylor Corse)

Cudworth *A Treatise Concerning Eternal and Immutable Morality* (edited by Sarah Hutton)

Descartes *Meditations on First Philosophy*, with selections from the *Objections and Replies* (edited with an introduction by John Cottingham)

Kant *The Metaphysics of Morals* (edited by Mary Gregor with an introduction by Roger Sullivan)

La Mettrie *Machine Man and Other Writings* (edited by Ann Thomson)

Leibniz *New Essays on Human Understanding* (edited by Peter Remnant and Jonathan Bennett)

Nietzsche *Human, All Too Human* (translated by R. J. Hollingdale with an introduction by Richard Schacht)

Schleiermacher *On Religion: Speeches to its Cultured Despisers* (edited by Richard Crouter)